Chemical Electrostatics

Fernando Galembeck • Thiago A.L. Burgo

Chemical Electrostatics

New Ideas on Electrostatic Charging: Mechanisms and Consequences

Springer

Fernando Galembeck
Institute of Chemistry
University of Campinas
Campinas, São Paulo, Brazil

Thiago A. L. Burgo
Departament of Physics
Federal University of Santa Maria
Santa Maria, Rio Grande do Sul, Brazil

ISBN 978-3-319-52373-6 ISBN 978-3-319-52374-3 (eBook)
DOI 10.1007/978-3-319-52374-3

Library of Congress Control Number: 2017935479

Printed on acid-free paper

This Springer imprint is published by Springer Nature
The registered company is Springer International Publishing AG
The registered company address is: Gewerbestrasse 11, 6330 Cham, Switzerland

This book is dedicated to
Emoke Hársi Galembeck, loving
and courageous companion on a long
journey.

Fernando Galembeck

and

To my mother and my father, for their love
and dedication.

Thiago A.L. Burgo

Preface

Electrostatics is now in its second decade of great activity that challenged well-established ideas substituting them for new concepts. As it often happens in science, this was made possible by new experimental tools like the scanning probe microscopes and scanning electrodes that produce 3D plots of charge, potential, capacitance, or conductance at solid and liquid surfaces. Another factor for change was the dissemination of other experimental tools like the noncontact electrostatic probes that measure electrostatic potential.

Researcher attitudes also changed and many decided to follow a suggestion made by Whitesides, accepting that macroscopic systems under equilibrium or quasi-equilibrium states may carry excess charge thus producing finite electric potential in their vicinity. The triboelectric series is another concept that was challenged, when macro- and microscopic potential maps showed that charge distribution on insulator surfaces is nonuniform, often following fractal patterns.

A major conceptual change is the recognition of the role of ions as charge carriers in a large number of important electrostatic phenomena. This is the result of many laboratory experiments and field results in different research disciplines, allowing researchers and engineers to benefit from a large amount of chemical knowledge to understand well-known but hitherto challenging phenomena. It is now possible to describe "space charge" in terms of ions and electrons, largely increasing the accuracy of the descriptions of the involved phenomena.

The atmosphere was acknowledged as the site of many phenomena producing electricity and this is related to water that assumes a paramount role in electrostatic phenomena. Water was previously seen as an agent for the dissipation of charge on electrified surfaces, due to its weak but non-negligible conductivity. Opposite to that, its ability to impart charge to solid surfaces is now recognized and it is extensively discussed in Chap. 6 of this book, on Hygroelectricity.

Triboelectricity is receiving a firmer mechanistic basis, since charge carriers in tribocharged polymers were identified as ionic species formed during mechanochemical reactions. Coupled to the demonstration that charge abatement on electrified polymer surfaces is also related to water adsorption, it can now be much better studied.

Another contribution came from acknowledging that many important metals are coated with a layer of oxide with low intrinsic conductivity. This also helped to understand interesting electrostatic phenomena in isolated metal objects.

The recent surge in the science of Electrostatics is thus due to new and often unexpected results that are being slowly examined, criticized, and absorbed by scientists and engineers. This new situation was not created by any bright new theory or by a single breakthrough that illuminated the way for scientists and engineers. For this reason, it is not easily captured by reading separate papers that contain a small but nevertheless essential part of the history. Moreover, there are a large number of papers and reports on various phenomena that were not followed up in the literature, remaining as valuable but uncorrelated and often forgotten pieces of information. These papers come from many different scientific disciplines, with different approaches, methodologies, and outlook.

This book is a systematic presentation of recent developments in Electrostatics, emphasizing those that are really new additions to the many excellent books that treat this subject in a more formal and abstract way. Recent developments are correlated to previously scattered experimental information that hitherto was considered too empirical to be considered in reference books. Recent and previous information is reported and discussed with an emphasis on physico-chemical and chemical events, whose elucidation is providing a better understanding of Electrostatics. This is why this book is named *Chemical Electrostatics.* Some readers may miss the abundance of mathematical equations often found in books and book chapters on Electrostatics. This is a deliberate choice of the authors, who did not want to repeat the contents of previous, excellent books.

Even though there is an emphasis on chemical and physico-chemical phenomena, chemical equations were kept to a minimum, acknowledging that chemical language is not familiar to many "hard" scientists and engineers.

Well-established concepts are used throughout the book, like pK and zeta potential, adsorption, electronegativity, electrical double layer, and many others that may not be familiar to some readers. The authors opted for not making specific introductions to them because this would largely lengthen the text. Moreover, these concepts are treated in detail in many reference books and in valuable materials accessible through the Internet. The authors strongly encourage readers to get well acquainted with those more basic concepts, whenever they feel they need it.

Chemistry made an invaluable contribution to our understanding of electrostatic phenomena and this will probably continue, for the foreseeable future. Chemistry is complex, the same as Nature. It is not amenable or properly explained by any encompassing theoretical approach; neither are most chemical phenomena predictable by any single theory. Given the importance of chemical phenomena to Electrostatics, the same applies to this discipline.

The authors expect that this phenomenological presentation will help researchers and engineers to devise new and better ways to reach some practical objectives. One is to increase human and property safety, avoiding undesirable consequences of electrostatic discharges, as discussed in Chap. 12. Chapter 13 presents new functional materials that can perform currently or previously unimaginable tasks, added to new products and processes to better human life, in a greener world.

Electrostatics has some appealing "green" features: large effects are produced using a few Joules, while great changes in the behavior of matter are obtained by adding or removing minute amounts of water. These features will allow Electrostatics to play a major role, in the quest for sustainability.

Campinas, São Paulo, Brazil — Fernando Galembeck
Santa Maria, Rio Grande do Sul, Brazil — Thiago A.L. Burgo

Acknowledgements

The authors are glad to acknowledge the support of:

Unicamp students whose work formed the original basis for most ideas and concepts developed in this book: André Herzog, Camila Rezende, Elizângela Linares, Érico Teixeira-Neto, Carlos Costa, Carlos Leite, Fábio Bragança, Leandra Santos, Leonardo Valadares, Lia Balestrin, Lucas Soares, Márcia Rippel, Rubia Gouveia, and Telma Ducati;

Pompeu Abreu-Filho, who organized the book chapters;

Maria do Carmo V. M. da Silva, who kept the laboratory on its feet while research was done;

The University of Campinas, where the major part of the work was done.

For many years, critical discussions were held with many persons. Their criticism and support is acknowledged but they are not to be criticized for any disputed ideas presented in this book: André Assis (Campinas), André Galembeck (Recife), Brian Vincent (Bristol), Daniel Lacks (Cleveland, OH), Darrell Velegol (University Park, PA), David Waddington (York), Gerald Pollack (Seattle, WA), Kazue Kurihara (Sendai), Oswaldo Alves (Campinas), Ponisseril Somasundaran (New York), Tom Healy (Melbourne).

Funds for equipment, fellowships, and laboratory running expenses were obtained from the Brazilian National Research Council (CNPq/MCTIC) and São Paulo State Research Foundation (Fapesp), who supported the Inomat project within the INCT (National Institutes for Science and Technology) program. Project numbers are: CNPq 573644/2008-0 and Fapesp 2008/57867-8. Earlier support from CAPES (fellowships), CNPq (Millenium Institutes Program) and Fapesp (Projeto Temático) is also acknowledged.

Excellent access to databases, journals, and other primary information materials was possible thanks to the *Portal de Periódicos* funded by Capes Foundation, from the Brazilian Ministry of Education.

Contents

Chapter 1
Living in an Electrified Environment

Contents

1.1 The Earth Capacitor

The Earth surface and the ionosphere are oppositely electrified, forming a huge capacitor [1]. The potential difference between the plates is in the 100–300 kV range and the average electric field at the Earth surface is 100 V/m [2].

This places all the natural and anthropic phenomena on Earth surface and atmosphere within an electrified environment, even though this is not easily perceived during fair weather. However, variable currents reaching a million amperes flow in the high-latitude auroral zones, near 100-km altitude inducing large currents within power and communication lines as well as on the ground and oceans that form the negative electrode in the Earth capacitor.

Electrical storms that are continuously taking place somewhere on Earth reveal the existence of huge amounts of electricity, producing spectacular displays.

F. Galembeck, T.A.L. Burgo, *Chemical Electrostatics*,
DOI 10.1007/978-3-319-52374-3_1

These are impressive for any person and many mythologies gave high status to the gods that master lightning.

The subject of atmospheric electricity has received much attention in the literature but it still poses great scientific challenges. Many books were written on this topic but there is not yet consensus on the mechanism for charge build-up and dissipation in the atmosphere.

The apparent absence of electricity in the atmosphere in fair weather is part of a feature of electrostatic phenomena that will be referred to in this book but in other contexts: charge on dielectrics produces few observable effects, provided the amount of charge remains below the threshold for discharge. Moreover, the low electric conductivity of the air and most non-metallic materials prevents the appearance of physiologically significant electric current, even under fields as high as 100 V/m. Since most effects of electricity are mediated by current, the significant electric potentials and fields may often go unnoticed until the dielectric undergoes ionization due to dielectric rupture or to impinging high-energy particles or still to charged particle motion following a Coulomb explosion.

However, there is a quiescent, low-density current in non-storm areas, opposite to the large but localized lightning current from the atmosphere to the Earth surface. This current is modulated by solar activity [3], mediated by cosmic ray ionization that is the major source of electrical conductivity in the air over the oceans and well above the continents. The solar wind modulates the flux of galactic cosmic rays impinging on Earth inversely with solar activity.

Assuming that the Earth capacitor is formed by two concentric spheres (ground and ionosphere) separated by 60 km, considering the Earth surface radius equal to 6300 km, its capacitance is very small, 12 nC and the stored energy amounts to 390 J only, notwithstanding the large potential difference between the capacitor plates. This cannot account for the devastating electric discharges produced during thunderstorms, showing that other sources of electricity are responsible for the powerful electric phenomena observed in Nature.

For this reason, independent of the size of the earth capacitor and of the impressive figures on electric potential difference between its plates, the onset of electrostatic discharge in the atmosphere depends also on phenomena leading to charge separation in clouds and in other aerosols, including those formed by ocean waves and waterfalls, organic chemicals and smoke.

1.2 The Global Atmospheric Electrical Circuit

Electric activity in between the ionosphere and the Earth surface is represented by the global atmospheric electric circuit shown schematically in Fig. 1.1 [4]. Thunderstorms and electrified rain clouds drive a DC current (~1 kA) carried by molecular cluster ions, while lightning phenomena drive the AC component of the global circuit. The Earth's near-surface conductivity ranges from 10^{-7} S m^{-1} (for poorly conducting rocks) to 10^{-2} S m^{-1} (for clay or wet limestone), to an average 3.2 S m^{-1}

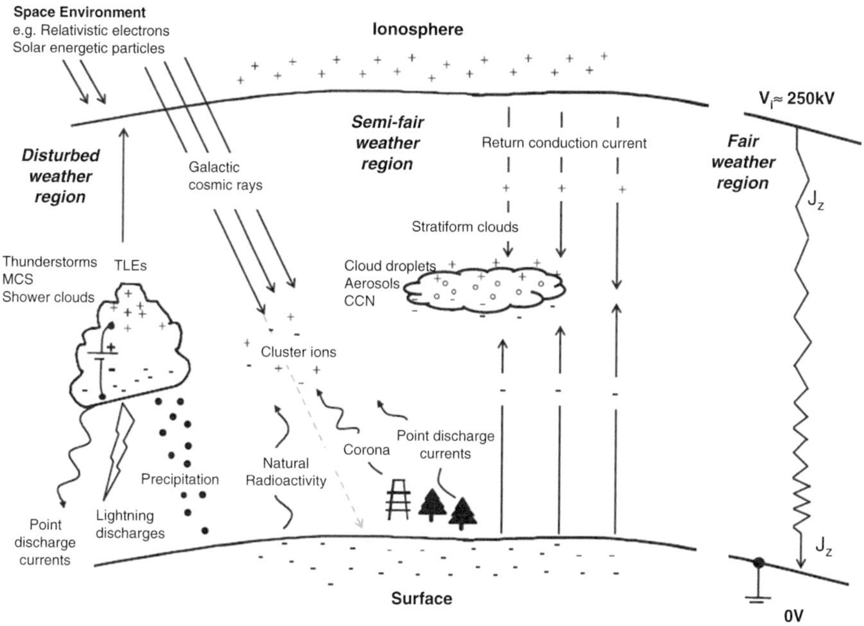

Fig. 1.1 Diagram showing various charge separation phenomena and electric currents that are part of the global electric circuit. Charge separation in thunderstorms creates a substantial potential difference within the atmosphere, while a small current flows vertically in fair weather and semi-fair weather regions. (MCS stands for Mesospheric Convective Systems, which are large thunderstorms; they produce sprites, Transient Luminous Events (TLEs); CCN stands for Cloud Condensation Nuclei). Reprinted with permission from [4]

for the ocean. Air conductivity inside a thundercloud, and in fair weather regions, varies from $\sim 10^{-14}$ S m^{-1} just above the surface to 10^{-7} S m^{-1} in the ionosphere at ~80 km altitude, depending on location pollution and height. Measurements have never shown a complete absence of fair weather electric field, suggesting that thunderstorm and other generators are operating continuously in maintaining the current flowing in the global circuit [5]. Nevertheless, the lifetime of electric energy in the atmosphere is very small—from about 10 to about 100 s [6]. In the current source regions, point discharge (coronal) currents play an important role below electrified clouds; the solar wind-magnetosphere dynamo and the unipolar dynamo due to the terrestrial rotating dipole moment also apply atmospheric potential differences [7].

Detailed measurements made near the Earth's surface show that Ohm's law relates the vertical electric field and current density to air conductivity. Stratospheric balloon measurements launched from Antarctica confirm that the downward current density is ~1 pA m^{-2} under fair weather conditions but solar energetic particles change the atmospheric conductivity and electric fields markedly [7]. Cloud-to-ground (CG) lightning discharges make only a small contribution to the ionospheric potential and sprites (namely, upward lightning above energetic thunderstorms) only affect the global circuit in a minuscule way.

Beyond its intrinsic interest, his topic has received further attention due to the emerging evidence for links among clouds, global temperatures, the global atmospheric electrical circuit, and cosmic ray ionization [8]. It is especially important to realize that a single measurement of Schumann Resonance, the Earth-ionosphere cavity resonance, has the potential of replacing the measurements of temperature all over the globe.

1.3 Electricity Produced Within the Earth Capacitor

The dielectric separating the plates of the Earth capacitor is neither uniform nor stable: quite the opposite, a number of mass transfer, phase change and chemical phenomena take place in its interior, often coupled to charge separation and segregation. These take place in much smaller volumes than the overall atmosphere but they produce powerful local effects.

1.3.1 Local Current Transients Within the Earth Capacitor: Lightning

Lightning [9] is one of the most familiar and widely recognized natural phenomena but it remains relatively poorly understood. Even basic questions of how charges are separated within thunderclouds, how lightning is initiated, and how it propagates for many tens of kilometers do not have widely accepted answers. Progress has been achieved thanks to advances in instrumentation, remote sensing methods, and rocket-triggered lightning experiments that are now providing new insights into the basic mass and charge transfer events leading to lightning. The recent association of intense bursts of X-rays and gamma-rays [10] with thunderstorms and lightning illustrate the new and interesting findings made in our atmosphere, providing a "plethora of challenging unsolved problems" [11] addressed in a number of reviews, papers, and books.

Lightning strikes the Earth 50–100 times each second and it causes heavy losses: the death of hundreds of people each year, forest fires, electric power outages, and damage to communications and computer equipment. Currently, a major concern of aircraft building companies is to preserve their resistance to lightning while replacing metal body parts with composite materials that offer significant gains in weight [12]. Lightning contributes to the production of nitrogen oxides in the atmosphere and this is an important natural process of nitrogen fixation that is essential for plant nutrition. This is remarkable, since the standard free energies of formation of NO (g) and NO_2 (g), $\Delta_f G°$, are respectively +87.6 and +51.3 kJ mol^{-1} showing that the formation of these two oxides from the elemental atmospheric N_2 and O_2 is non-spontaneous. It nevertheless takes place, although at a limited extent, when triggered by lightning.

Intense electric fields are observed during thunderstorms and lightning takes place when the thunderstorm electric field exceeds a critical value, called the breakeven field [13]. It is remarkable that at any altitude the breakeven field is lower than the field usually presumed to be required for dielectric breakdown or for streamer propagation. Electric fields associated with thunderstorms can cause a person's hair to stand on end and they also produce more serious events, like corona discharges from antennas, trees, bushes, grasses, and sharp objects. They also have the ability to uplift either charged ice particles containing microorganisms [14] or free microorganisms, which inherently contain an electric charge, into the stratosphere at altitudes of 40 km and above. This shows that the space in between the Earth capacitor plates is not only under a significant average electric field but higher endogenous fields are created in its interior at a high frequency.

Several processes acting below, in and above thunderstorms and in electrified shower clouds drive currents that are part of the global atmospheric electric circuit. Recent simulation work [15] shows that moderate negative cloud-to-ground lightning from the base of a thunderstorm increases the ionospheric potential above the thundercloud by 0.0013% that changes the ionosphere surface equipotential, also by 0.0013%. A similar but positive cloud-to-ground lightning discharge decreases the ionospheric potential by 0.014%. Given the global average rate of lightning discharges and the relevant relaxation times, it is found that negative and positive discharges change the ionospheric potential by only a few percent contributing only ~1% to maintaining the high potential of the ionosphere. Thus, the net upward current to the ionosphere due to lightning is only ~20 A. On the other hand, it is concluded that conduction and convection currents associated with power generation within thunderclouds and electrified shower clouds contribute essentially equally (~500 A each) to maintaining the ionospheric potential.

1.3.1.1 Dry Lightning

Lightning is usually associated with thunderstorms and thus to atmospheric water. However, it is also observed in dry environments where water can hardly play an important role.

Two relevant cases are lightning associated with volcano eruptions and sand storms in the desert.

1.3.1.2 Lightning and Volcanic Activity

Intense electrical activity and lightning is often associated with explosive volcanic eruptions. This has been known for a long time but the mechanism for charge separation and violent discharge are not yet fully understood [16].

Likely processes that have been proposed to explain the electrification of volcanic plumes are quenching magma-water interactions, particle breakdown and friction of fine ash and the freezing of plume water at height. The difficulties to obtain

Fig. 1.2 Plume lightning observed during the 2010 Eyjafjallajökull volcanic eruption. Image copied from the USGS site [18]

measurements during volcanic lightning have prevented progress in sorting out the relevant mechanisms but the Eyjafjallajökull volcanic eruption (Fig. 1.2) in Iceland in April–May 2010 made a great contribution in this direction. The eruption lasted 39 days, going through different phases while the surrounding atmosphere also changed. A major change was observed on 11 May, when intense lightning activity started to last for 10 days, following a 7-day period of little activity. The increased lightning intensity was concurrent with a change in the conditions of the ambient atmosphere, bringing down the altitude of the isotherms for droplet freezing (about −20 °C, below the plume top. Lightning was then associated with the existence of solid water particles in the surrounding atmosphere, analogous to processes in meteorological thunderclouds.

The difficulty to make field measurements led researchers to create an experimental model for this phenomenon [17], by video recording rapid decompression experiments of gas-particle mixtures under controlled conditions, using a high-speed camera and two antennas. The lightning is controlled by the dynamics of the jet and by the abundance of fine particles in the jet, whose relative motion creates the required electric field. These findings suggest that electrostatic monitoring may contribute to the forecasting of volcanic ash emissions.

1.3.1.3 Lightning and Sandstorms

"Dust devils" are swirling tall columns of sand dust observed is dry areas. Field reports on potential gradients observed during the displacement of dust devils in New Mexico and in the Sahara were made by Freier [19] and Crozier [20] respectively, who measured potential gradient depressions reaching 540 V/m. The authors attempted to devise an electrical structure for the dust devil but they acknowledged limitations due to the lack of a model for charge generation in the dust.

Dust devils were also observed on Mars surface, as "a monster column towering kilometers high and hundreds of meters wide, 10 times larger than any tornado on Earth" [21]. Mars' dust devils are electrified, producing incessant crackling, flashing and electromagnetic interference from lightning to the point of "arcing your spacesuit or vehicle and creating magnetic interference". For these reasons, dust storm electrification is listed among major risks to Martian missions [22].

The field observed during strong storms was observed as generally opposite to the average fair-weather field reaching values as high as 20 kV/m, 1 m above ground [23]. Space charge as high as 1.6×10^{-6}–1.6×10^{-7} C m^{-3} of either polarity were measured 1.25 m above ground in the United States [24] and in Sahara.

In these and related studies the amount and sign of charge shows a correlation to particle size that has been verified in laboratory studies [25] and will be discussed in Chaps. 8 and 9, in this book.

Sand electrification based on size-dependent particle charging is now a source of inspiration for energy harvesting [26].

1.3.2 Coupling to Earthquakes

Recently, the electromagnetic coupling between seismic activity and the ionosphere was considered within the framework of the Global Electric Circuit concept. Anomalous variations in the ionosphere were associated with the earthquake preparation process, their temporal and spatial characteristics. Changes in the troposphere lead to sharp and fast changes of atmospheric parameters including the electric properties of the boundary layer and contributing to seismo-ionospheric coupling [27].

1.3.3 In Other Planets

Atmospheric electrification is not restricted to the Earth, as already shown in the section on "dust devils". Electrification is a fundamental process in planetary atmospheres, found widely in the solar system [28]. It is most evident through lightning discharges, which can influence the chemical composition of the atmosphere. Detailed modeling of the formation of intense electric fields in Mars led to the conclusion that lightning should take place in this planet [29] and this was indeed later verified [30]. Electrification also affects the physical behavior of aerosols and cloud droplets that in turn modify the absorption and scattering of radiation so as to determine an atmosphere radiative balance. Since lightning produces complex molecules through thermodynamically non-spontaneous reactions, it can contribute to the appearance of otherwise unexpected chemicals, including prebiotic species [31].

1.4 Electricity in the Crust: The Self-Potential

The *self-potential* measured between two electrodes contacting different points in the ground is an important topic for researchers from many areas, especially geophysics [32, 33], hydrology, archaeology [34], and volcanology. It is a passive method based on the natural occurrence of electrical fields on the Earth's surface. In some cases, the self-potential is largely due to electrokinetic phenomena [35] but thermoelectric [36] and electrochemical [37] contributions are also acknowledged in other cases. It contributes to the localization and quantification of the flow of groundwater as well as pollutant plume spreading and to the estimation of pertinent hydraulic properties of aquifers [38].

The observed potential differences are in the mV to V range, much lower than atmospheric potentials. On the other hand, they are often measured in low-resistivity moist porous media.

1.5 Human Perception of Environmental Electricity

Human senses do not perceive electricity in any way close to their perception of light, heat and cold, vibrations, mechanical stresses, and chemicals. Just looking at a power transmission line we cannot tell that if it is electrified or not, even though some persons can identify the ozone smell due to large electric fields. Birds are also insensitive to electrified objects and many die, when they touch two cables at the same time. However, some displays of light and sound caused by electric discharge are so powerful that humans associated lightning and thunder to the most powerful gods in their mythologies, from Zeus and Jupiter in ancient Greece and Rome to Tupan, among the tupi-guaranys. Contemporary life is filled with electrical apparatuses but humans continue to be insensitive to the surrounding electric fields and even to electric potential, although we respond to electric current. This is intriguing: why aren't we humans sensitive to electric fields, as much as we are sensitive to electromagnetic waves, to temperature, mechanical contact, pressure, and other types of macroscopic physical agents?

Perception of electricity by humans is often described as "electric shock" that is unpleasant and may lead to the loss of consciousness and even to death. The situation is similar for other living beings: a tree under a thunderstorm does not show any visible response to the surrounding electric fields, while it is being strongly deformed by the wind. However, one instant later it may be hit by lightning that will put it ablaze. Human sensing of electricity often follows a step function: no perception almost every time but eventually overwhelming or even fatal, for a short time. Year after year, lightning kills many persons who remained under heavily clouded skies: everybody knows the risk but the lack of sensitivity suggests that nothing will happen, until lightning happens.

Some signs or the surrounding electricity are perceived, at times: pants stick to our legs, hair rises up, and fur bends, but these events do not have the same regularity and frequency of our perception of light, heat, or sound.

On the other hand, negative effects of electric fields on human health are often mentioned, associating the incidence of cancer and the exposure of electric fields in the vicinity of power transmission lines. There are laboratory reports on the effect of endogenous and exogenous electric fields [39, 40] on living tissue and biological materials. Moreover, electrokinetic techniques have been used [41] in tissue engineering. However, systematic critical examination of electric fields on biological systems shows that electric fields are less damaging than magnetic fields [42].

Lack of perception of electric fields is widespread among living beings, with a few notable exceptions like the "electric fish" and other aquatic species [43], bees [44], cockroaches, and few others.

Humans and most other living species are insensitive because they are electrically shielded. This is essential for survival, because life is extremely dependent from electric phenomena for perception, information, decision-taking and motion, either automatic or under brain control. The relevant potential differences in our muscles, nerves, and brain are well below 1 V and fortunately they are shielded from the 100 volts per meter gradient in our surrounding atmosphere. Human electrical activity is detected in many ways, as electrocardiograms, brain wave, and other electric signals that require sophisticated instrumentation, but they are not consciously perceived by other persons.

Human body shielding is done by water and it is based on water ability to acquire net charge, countering low-frequency or static fields. The physical–chemical basis for this impressive role of water is described in this book for the first time in a comprehensive way, in Chap. 6.

The limited perception of electric fields by humans leads many persons to believe that we live in an electroneutral world, exempt of electric phenomena. Quite the opposite, we live in an electrified environment created by powerful external and internal factors.

1.6 Conclusions

The Earth surface and its atmosphere are electrified environments due to the permanently charged capacitor formed by the ionosphere and the crust surface, added to a number of natural phenomena producing large and small local electric fields, like thunderstorms, inter-particle collisions, and tectonic activity. The latter provoke changes with relaxation times in the range of minutes to hours, while the Earth capacitor changes following solar activity and Earth circadian and annual periods.

None of these changes is perceived by human senses, going largely unnoticed until electric discharge takes place, often with severe material and personal consequences.

Recognizing that mutable electric patterns are an intrinsic feature of our environment will probably allow every person, especially researchers and engineers, to acquire a better understanding of natural and anthropic phenomena.

The recent findings in this area will probably contribute to two important sets of emerging technologies: energy harvesting and environmental engineering.

References

1. Geophysics Study Committee, Overview and Recommendations (1986) In: The Earth's electrical environment. National Academies, Washington, DC, p 120. http://www.nap.edu/catalog/898/the-earths-electrical-environment. Accessed 16 Jun 2016
2. Krehbiel PR (2016) The electrical structure of thunderstorms. In: The Earth electrical environment, Chapter 8, p 90. http://www.nap.edu/catalog/898/the-earths-electrical-environment. Accessed 16 Jun 2016
3. Harrison RG, Usoskin I (2010) Solar modulation in surface atmospheric electricity. J Atmos Sol Terr Phys 72:176–182
4. Rycroft MJ, Nicoll KA, Aplin KL, Harrison RG (2012) Recent advances in global electric circuit coupling between the space environment and the troposphere. J Atmos Sol Terr Phys 90-9:198–211
5. Siingh D, Gopalakrishman V (2007) The atmospheric global electric circuit: an overview. Atmos Res 84(2):91–110
6. Mareev EA, Anisimov SV (2009) Lifetime of the thunderstorm electric energy in the global atmospheric circuit and thunderstorm energy characteristics. Atmos Res 91:161–164
7. Rycroft MJ, Harrison RG et al (2008) An overview of earth's global electric circuit and atmospheric conductivity. Space Sci Rev 137:83–105
8. Harrison RG (2004) The global atmosphere electrical circuit and climate. Surv Geophys 25:441–484
9. Uman MA (1994) Natural lightning. IEEE Trans Ind Appl 30(3):785–790
10. Williams ER (2005) Lightning and climate: a review. Atmos Res 76:272–287
11. Dwyer JR, Uman MA (2014) The physics of lightning. Phys Rep 534:147–241
12. Laroche P, Blanchet P, Dellanoy A, Isaac F (2012) Experimental studies of lightning strikes to aircraft. AerospaceLab:1–13. http://www.aerospacelab-journal.org/sites/www.aerospacelab-journal.org/files/AL05-06_0.pdf. Accessed 16 Jun 2016
13. Marshall TC, McCarthy MP (1995) Electric field magnitudes and lightning initiation in thunderstorms. J Geophys Res 100:7097–7103
14. Dehel T, Lorge F et al (2008) Uplift of microorganisms by electric fields above thunderstorms. J Electrostat 66:463–466
15. Rycroft MJ, Odzimek A et al (2007) New model simulations of the global atmospheric electric circuit driven by thunderstorms and electrified shower clouds: the roles of lightning and sprites. J Atmos Sol Terr Phys 69(17–18):2485–2509
16. Pähtz T, Herrmann HJ, Shinbrot T (2010) Why do particle clouds generate electric charges? Nat Phys 6:364–368
17. Cimarelli C, Alatorre-Ibargüengoitia MA et al (2014) Experimental generation of volcanic lightning. Geology 42:79–82
18. Eyjafjallajokoll plume lightning. https://volcanoes.usgs.gov/volcanic_ash/lightning.html. Accessed 21 Oct 2016

19. Freier GD (1960) The electric field of a large dust devil. J Geophys Res 65(10):3504
20. Crozier WD (1964) The electric field of a New Mexico dust devil. J Geophys Res 69(24):5427
21. Bell T (2005) Devils of Mars. http://science.nasa.gov/science-news/science-at-nasa/2005/14jul_dustdevils. Accessed 16 Jun 2016
22. Beaty DW, Snook K et al (2005) An analysis of the precursor measurements of mars needed to reduce the risk of the first human missions to Mars. Unpublished white paper, 1–77, posted June, 2005 by the Mars Exploration Program Analysis Group (MEPAG). http://mepag.jpl.nasa.gov/reports/cfm. Accessed 16 Jun 2016
23. Stow CD (1969) Dust and sand storm electrification. Weather 24:134–140
24. Kamra AK (1972) Measurements of the electrical properties of dust storms. J Geophys Res 77(30):5856–5869
25. Bilici MA, Toth JR III et al (2014) Particle size effects in particle-particle triboelectric charging studied with an integrated fluidized bed and electrostatic separator system. Rev Sci Instrum 85:103903
26. Hu W (2016) Wind-blown sand electrification inspired triboelectric energy harvesting based on homogeneous inorganic materials contact: a theoretical study and prediction. Sci Rep 6:19912
27. Pulinets S, Davidenko D (2014) Ionospheric precursors of earthquakes and global electric circuit. Adv Spec Res 53:709–723
28. Yair Y, Fischer G et al (2008) Updated review of planetary atmosperic electricity. Space Sci Rev 137:29–49
29. Melnik O, Parrot M (1998) Electrostatic discharge in Martian dust storms. J Geophys Res 103(A12):29107–29117
30. Steigerwald B (2006) Electric dust storms on Mars. http://www.nasa.gov/vision/universe/solarsystem/mars_soil_chem.html. Accessed 16 Jun 2016
31. Harrison RG, Aplin KL et al (2008) Planetary atmospheric electricity. Space Sci Rev 137:5–10
32. Jouniaux L, Maineult A et al (2009) Review of self-potential methods in hydrogeophysics. Comptes Rendus Geosci 341:928–936
33. Nyquist JE, Corry EC (2002) Self-potential: the ugly duckling of environmental geophysics. Leading Edge 21(5):446–451. http://tle.geoscienceworld.org/. Accessed 16 Jun 2016
34. Aspinall A, Gaffney CF (2001) The Schlumberger array: potential and pitfalls in archaeological prospection. Archaeol Prospect 8:199–209
35. Moore JR, Glaser SD (2007) Self-potential observations during hydraulic fracturing. J Geophys Res 112:B02204
36. Revil A, Karaoulis M et al (2013) Thermoelectric self-potential and resistivity data localize the burning front of underground coal fires. Geophysics 78(5):B259–B273
37. Revil A, Ethouarne L, Thyreault E (2001) Tomography of self-potential anomalies of electrochemical nature. Geophys Res Lett 28(23):4363–4366
38. Corwin RF, Hoover DB (1979) The self-potential method in geothermal exploration. Geophysics 44:226–245
39. Anastassiou CA, Montgomery SM, Barahona M, Buzsaki G, Koch C (2010) The effect of spatially inhomogeneous extracellular electric fields on neurons. J Neurosci 30:1925–1936
40. Fröhlich F, McCormick DA (2010) Endogenous electric fields may guide neocortical network activity. Neuron 67:129–143
41. Markx GH (2008) The use of electric fields in tissue engineering: a review. Organogenesis 4:11–17
42. Kheifets L, Renew D, Sias G, Swanson J (2010) Extremely low frequency electric fields and cancer: assessing the evidence. Bioelectromagnetics 31:89–101
43. Alves-Gomes J (2001) The evolution of electroreception and bioelectrogenesis in teleost fish: a phylogenetic perspective. J Fish Biol 58(6):1489–1511
44. Greggers U, Koch G, Schmidt V, Dürr A, Floriou-Servou A, Piepenbrock D, Göpfert MC, Menzel R (2013) Reception and learning of electric fields. Proc R Soc B 280(1759):1471–2954

Chapter 2
Electroneutrality: When and Where?

Contents

2.1 A Widespread Belief

Many persons educated in science and engineering tend to view their surroundings as being electrically neutral, except for the electricity used to power equipment and lighting ware in any anthropic environment. When asked about electricity in a quiescent environment, scientists and engineers often invoke electroneutrality, including an "Electroneutrality Principle" that will be discussed ahead.

This is not surprising, since unbalanced electrostatic charge is hardly conspicuous: it is not perceived by our senses, it does not provoke frequent sound emission or color change and some more frequent visual effects are too subtle to be noticed,

F. Galembeck, T.A.L. Burgo, *Chemical Electrostatics*,
DOI 10.1007/978-3-319-52374-3_2

Fig. 2.1 Electrical deflection of a distilled water droplet stream within the electric field between two aluminum deflector plates separated by 2 cm Styrofoam spacers. The plate on left was grounded and a potential was induced on the right-hand plate by a statically electrified balloon that was outside the field of view to the right. Reprinted with permission from [1]

as for instance in the case of water drop deformation due to excess charge. This will be treated in detail in Chap. 6, but a short description follows.

Water drops are deformed in the presence of an electrified body and the trajectory of falling water is modified, in the presence of an electrified object, as shown in Fig. 2.1. This has been known for more than two centuries and it is used as a demonstration of molecular polarity in textbooks. Unfortunately, this is eventually accompanied by incorrect explanations, as discussed by Ziaei-Moayyed and Goodman [1]. Some related movies that can be seen in the Internet are supplemented by comments and explanations that are examples of misconceptions on electrostatic phenomena. A microfluidic version of this experiment accompanied by a quantitative analysis was recently published [2], see Fig. 2.2.

2.2 Charge Accumulation, Electrostatic Discharge

Electrostatic charge accumulation usually goes unnoticed, until a critical value is reached for the electric field and a discharge takes place, often leading to fire, explosions, and personal injuries. This kind of behavior is quite different from many other types of physical phenomena. For instance, an object undergoing mechanical stress, torsion, or compression normally shows some deformation, prior to breaking or acquiring a new shape due to plastic deformation. When water is heated, its

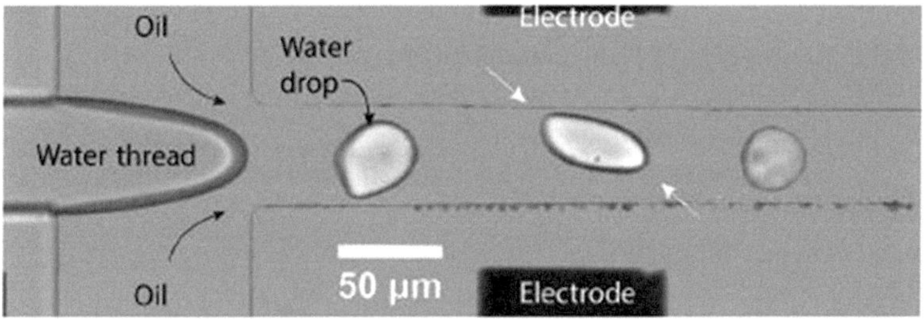

Fig. 2.2 Water droplets in microchannels as they pass between two electrodes. Droplet shape becomes elliptical and they tend to emit daughter droplets, as in the classical Rayleigh explosions. Reprinted with permission from [2]

temperature increases before it starts to boil. On the other hand, unwanted electrostatic discharges take place without any warning: most systems will build-up electric charge that goes unnoticed, until the limit is reached.

A popular class demonstration experiment is the "Kelvin dropper", an ingenious but simple experimental set-up where falling water drops produce sufficient electricity to provoke sparks. Most reports of this experiment are rather qualitative since voltage build-up is not usually measured, but it has recently been used in microfluidic devices [3] with attractive features, including the possibility of energy scavenging.

Rather dangerous, electrical discharge may be triggered just by making some geometrical change in the arrangement of solids and liquids, for instance by bringing a grounded conductor closer to a pile of charged dielectric powder. The electric field may increase to the point of provoking dielectric rupture of the air and thus provoking a discharge. So, performing what may look like a safety procedure may indeed trigger an explosion. The situation may become still more complex if this is coupled to motion relative to a magnet that will also provoke charge displacement [4].

2.3 Electric Potential, Electric Field, Electrochemical Potential

A complicating factor in the understanding of electrostatic phenomena is the mixed use of electric field or electric potential to describe electric effects on various phenomena. Electric field fits well within mechanical reasoning, since it is easily associated with forces acting on charges and dipoles and thus it is easily related to the motion of charge carriers or the orientation of objects containing separated charges, as dipoles of multipoles.

On the other hand, electric potential fits easier within thermodynamics and its use is convenient, as an intensive factor coupled to charge, which is extensive. It was used by Tolman, in his outstanding work, e.g. on the determination of electrochemical transference numbers of ions by centrifugation. In his writing,

"The method consists in the measurement of the electromotive force produced between electrodes placed at the central and peripheral ends of a rotating tube containing the electrolyte. An equation can be derived, connecting this electromotive force and the transference number of the salt with the speed of rotation, the density of the solution, and the molecular weight and the "partial" specific volume of the substances involved..." [5].

The existence of sedimentation potentials, by itself, is a powerful demonstration that electroneutrality is not expected along any solution or dispersion contained within a tube that underwent centrifugation. Simple extension of Tolman's equations to larger solutes or particles dispersed in a liquid shows that it should not also be expected along the depth of any water body where charged particles (clays, sand, humic matter) are dispersed, settling under gravity. So, electric potential gradients are expected even under sedimentation equilibrium conditions, in aqueous electrolytes.

Electrochemical potential is currently widely used by two groups of researchers: electrochemical kineticists and electrophysiologists. Both groups are well aware of the importance of the electrochemical potential in the treatment and understanding of the phenomena they are interested in: electrochemical kinetics makes explicit use of the effect of local electric potential on the activation energy of an electrode reaction, while the importance of electric phenomena in signal transmission in living cells and tissue is now perceived even by the layman using medication related to sodium pumps and other nanobiological gear. However, they are a minority. This is verified by comparing the number of entries in the Web of Science, over all years recorded, at a given time: entries for "chemical potential" number 14,180 vs. 2852 for "electrochemical potential".

Instrumentation researchers have been familiar with non-uniformity of electric potential in insulator surfaces. For instance, Lion stated that "The (usually irregularly distributed) electrostatic charges on the surface of an insulator can be detected and measured by the use of a moving electrode" [6]. His recognition of charge non-uniformity was not widespread and it is usually ignored by experimenters using plastic labware.

2.4 Taking Electroneutrality for Granted

Examples of researchers treating excess charge in matter as a non-issue can be taken from any scientific or technological area: for instance, polymer textbooks only refer to electric charge in the chapters explicitly related to ionic species, either on polyelectrolytes or living polymerization initiated or involving ionic species, notwithstanding the obvious and frequent occurrence of excess charge in polymer materials exiting mills and processing equipment.

Charge build-up during operations widely used in the chemical, pharmaceutical, food, and related industries has been reported. Particles are often electrostatically charged by frictional contact during powder-handling operations [7]. More generally, during the operation of multiphase systems such as fluidized beds, electrostatic charges are generated when the materials involved are dielectric in nature [8]. The same is observed during a melt agglomeration process [9] and in many other cases.

On the other hand, recent work from the author's laboratory shows that water is often non-electroneutral and that it contributes in many ways to impart non-electroneutrality to other substances [9, 10] that will be dealt in a separate chapter.

2.5 The Electroneutrality Principle

This principle is not often mentioned in the literature. A search in Web of Science using "electroneutrality principle" as the entry returned only 58 documents and in many cases electroneutrality is taken as a rule or principle, without further analysis. For instance, the statements "The main physicochemical principle which must be accomplished in body fluids, is the rule of electroneutrality" [11] and "The mathematical model is based on principles of mass conservation, acid-base equilibria, and electroneutrality" [12] are read in the recent literature.

This principle is an entry in IUPAC Gold Book [13] where we can read, under "electroneutrality principle": "The principle expresses the fact that all pure substances carry a net charge of zero." This is correct for "pure substances" but strictly under some assumptions, e.g. zero electric potential [14]. Under non-zero potential, satisfying the condition of minimum electrochemical potential requires that any system contains excess cations under negative potential and excess anions under a positive potential.

The assumption of electroneutrality appears explicitly in recent scholar texts, e.g.: "Ions are not independent in ionic solids, where we take for granted the fact that there are exactly equal numbers of Na^+ and Cl^- ions (or we would be electrocuted each time we salt our food)…" and "One *always without exception* dissolves a strictly neutral salt in water to make an ionic solution…no violation of electrical neutrality is significant in chemical units…" [15]. The electrocution argument is certainly not correct, because this implies the onset of electric current but this is not to be expected from the immobile crystal ions. Moreover, given the size of Avogadro's number, having exactly equal numbers of Na^+ and Cl^- ions implies that no imbalance takes place within a population in the range of 10^{23} individuals, what is not to be expected.

A related issue is the widespread assumption of non-volatility of ionic species, from room-temperature liquids. This matter was discussed in detail by Rockwood [16], connecting it to thermionic emission, and recalling that ions may undergo thermionic emission [17]. The same physics developed by Fermi for the treatment of an electron gas in equilibrium with a metal must apply to ions as well as to electrons, which implies that ions must also have a vapor pressure that is usually neglected. Evidence in favor of ion transfer across solid- and liquid-gas interfaces will be presented, in various chapters of this book, especially Chaps. 4 and 6, on Charge Mobility and Hygroelectricity, respectively.

2.6 Pauling's Principle of Electroneutrality

Pauling's principle of electroneutrality played an important role in the study of molecular structure. The original statement by Pauling in 1948 says that each atom in a stable substance has a charge close to zero and it appears in the Dover reprinting of the 3rd edition of his "General Chemistry" book as follows: "Stable molecules and crystals have electronic structures such that the electric charge of each atom is close to zero. Close to zero means between -1 and $+1$" [18]. This principle has been used to predict relative stability of molecular resonance structures, to explain the stability of inorganic complexes, π-bonding in compounds and polyatomic anions. However, modern computational techniques may indicate greater ionic character than predicted by this principle [19].

To the present authors, it is not clear how and why this statement may lead anyone to believe that macroscopic matter is also electroneutral. It is well known that the collective behavior of molecules is often not predictable just considering the properties of individual molecules. Moreover, the advent of nanotechnology made clear that size matters and many properties of a given substance may change with size. For these reasons, Pauling's principle of electroneutrality is not a sound basis for assuming that matter is electroneutral, notwithstanding its importance in the determination of molecular structure and properties.

2.7 Factors of Non-Electroneutrality

2.7.1 Dangling Bonds

Many scientists and engineers were taught to view common matter as if it were a more or less complex assembly of great numbers of well-behaved ions and molecules, in the Avogadro range. Some situations that do not fit within this picture are usually overlooked. For instance, dangling bonds are structures not contained within the usual descriptions of the molecules and crystalline solids [20] and they have been reported in many cases: at the ends of graphene molecules [21]; in silicon [22], and germanium [23]. OH dangling bonds were detected in the hydration shells around dissolved nonpolar (hydrocarbon) groups [24]; in polymers, e.g. PTFE [25] where it accounts for ferromagnetism; in GaN [26].

Recognizing the existence of dangling bonds in rather simple systems that are among the most intensively studied by chemists, physicists, materials scientists, and engineers should prompt researchers to accept other departures from standard descriptions, including that there are not "exactly equal numbers of Na^+ and Cl^- ions" in NaCl crystals.

2.7.2 *Are Ionic Crystals Electroneutral?*

Crystalline solids themselves have a large number of possibilities for developing excess charge, positive or negative. A fundamental reason is entropy. It is well established that defects develop spontaneously on crystals, since they contribute to increase entropy and to decrease free energy. This often goes at the cost of adding energy to the crystal to create the defect.

A very simple case is a salt crystallizing from aqueous solution. If electrostatics were the only factor for ion deposition on the growing crystals, then perhaps each growing crystal would be exactly neutral at each point in time. However, a neutral salt crystal within a solution of its ions is a rare situation. This is abundantly evidenced by considering surface properties of salt crystals in the presence of dissolved ions. In this area, Lyklema and his collaborators in Wageningen and elsewhere produced a huge amount of rigorous information on excess charge on silver iodide crystals, determined either by potentiometric titration or more indirectly by zeta potential measurements. They examined adsorption and double layer capacitance [27] in the presence of various other ions, polymers [28] under variable pH, co-solvent and variable temperature [29] during five decades, creating solid ground for understanding the behavior of real ionic crystals in water.

Other important contributions come from different groups. For instance, "in the vicinity of the point of zero charge the Nernst equation accurately gives the surface charge" [30].

Indeed, the interaction of water with various ions at crystal surfaces cannot be expected to be non-specific. Silver iodide has also been studied in this respect due to its important role nucleating ice formation, in the atmosphere [31]. A recent finding is that "Water molecules strongly adsorb onto the Ag^+ terminated face to give a well-ordered hexagonal ice-like bilayer that then acts as a template for further ice growth", but water does not adsorb at I^- terminated basal face or the prism and normal faces [32]. A study of the nucleating activity of silver iodide in super-cooled water showed its dependence upon the potential-determining silver and iodide ions [33].

Apparently, all the knowledge referred to in previous paragraphs is left aside when students are taught to use electroneutrality in their calculations, indiscriminately. For instance, they learn that in equilibrium reactions like $AgI = Ag^+ + I^-$ the two concentrations are equal, $[Ag^+] = [I^-]$ and $K_w = [Ag^+]^2 = [I^-]^2$. However, under most conditions the AgI particles contain non-zero charge, evidenced by electrophoresis, titration data, and other data. Consequently, $[Ag^+] \neq [I^-]$.

2.7.3 *New Ion Sources for Mass Spectrometry*

Mass spectrometry is currently a large family of analytical techniques with a great diversity of equipment configurations, procedures, and areas of interest. The initial step in the analysis of neutral molecules by mass spectrometry is the production of molecular ions that formerly was done under high vacuum and this was an important limitation for mass spectrometry. In 1994, Hirabayashi and collaborators [34] revolutionized mass spectrometry by introducing an atmospheric pressure chemical ionization (APCI) technique termed "sonic spray ionization" (SSI) that "was unique and revolutionary because it introduced a new concept of ionization to mass spectrometry" producing ions without the assistance of voltage, radiation, or heating. The charged droplets were produced simply by spraying an acidified solution of the analyte in methanol at sonic speed and charge separation was assigned to a statistically unbalanced distribution of cations and anions. This opened the way to ambient mass spectrometry techniques like EASI that brought mass spectrometers into the "real world" for the following reasons: (1) its great simplicity, because only compressed nitrogen or air is required; (2) its ability to simultaneous produce both negatively and positively charged droplets, hence no need to switch high potentials in changing from EASI(+) to EASI(−); (3) the low charge concentration on the droplets, which seems to improve signal-to-noise ratios; (4) the extreme softness of the ionization process; (5) no thermal degradation. One limitation of EASI is the ultra-high-velocity spray stream, which can easily blow samples away, but the formation of bipolar aerosol using only low-pressure systems was recently disclosed [35] and it may find applications in MS, in the near future.

2.7.4 Contact Charging, Mechanochemistry, Tribochemistry

Contact charging is known to almost every person, it has been known since the ancient Greeks but is description in terms of the atomic-molecular theory as it was established nearly one century ago is still a big challenge [36]. Nevertheless, it shows that each one of two solids acquire excess charge upon contact, meaning that neither becomes electroneutral.

Beyond contacting other solids and liquids, any substances and materials in the real world are subject to handling, contact with other substances including gas and liquid flow over their surfaces, cutting and milling, tension and compression and a number of other kinds of mechanical actions. Persons experienced in different kinds of operations, in any environment, are familiar with the appearance of charged particles, liquids and solids, in many situations, from cement to pharmaceutical and food processing [37, 38].

Following IUPAC, the term "mechano-chemical" is used for a "Chemical reaction that is induced by the direct absorption of mechanical energy." The Gold Book also notes that "shearing, stretching, and grinding are typical methods for the mechano-chemical generation of reactive sites, usually macroradicals, in polymer chains that undergo mechano-chemical reactions" [13]. Judging by the frequent observation of electrostatic charging in mechanically processed materials, many mechano-chemical reaction products carry excess charge that provoke various effects in the making of various types of solids and powders [39].

However, electrostatic separation based on different tribo-electric charging behaviors of components has emerged as a novel, sustainable dry fractionation process [40].

Milling solids provide many good examples, since the resulting powders are often sufficiently charged to adhere to equipment surfaces, blocking its chambers and provoking the interruption of the operation [7]. Since industrial milling operations use up to 5% of the electricity produced in the world, and it dissipates hitherto unknown amounts of charge in the environment, perhaps this would be a good area for introducing energy harvesting equipment.

Unfortunately, mechanochemistry, tribochemistry, and even tribology at large are not popular areas among basic scientists, even though there are excellent reviews and books on this topic [41–46], bringing information, showing opportunities in chemical synthesis [47, 48] while aiming to create a unifying framework that enables predictions of force-induced reactivity [49, 50]. This topic will be treated in a separate chapter in this book and the message left at this point is this: common mechanical action on solids and complex liquids may cause significant chemical change, including the formation of high-energy species like free-radicals and ions that may soon react forming other high-energy species with half-lives reaching many months, in some cases. This is well represented by the words "triboplasma", coined by Heinicke and "magma-plasma", by Thiessen, to describe the unique chemical and physical environments created by mechanical action on different solids.

2.7.5 Liquid Junction Potential and Membrane Potential

Liquid junction potential (LJP) is a concept familiar to electrochemists, electrophysiologists [51], and to any person that uses a combined glass-reference electrode for pH measurement. It appears when a liquid–liquid junction is made, contacting two liquids with different concentration and chemical composition of electrolyte solutes [52], or even two different liquids [53].

It is easily understood considering that the diffusion coefficients of the cation and anion in an electrolytic solution are usually different, as well as their electrophoretic mobilities and transport numbers as shown in Fig. 2.3. The magnitude of junction potentials cannot be overlooked and it has been demonstrated that LJPs formed in microchannels can induce appreciable electrophoretic transport of charged species without the use of electrodes or an external power supply [54].

Membrane potential is observed whenever a membrane separates two electrolytic solutions. This is important in various natural and technological processes and it has been widely used in chemical analysis, in the glass electrodes [55] for pH and other ions measurement and various membrane electrodes that were developed for a large number of analytes [56]. Moreover, it is a basic concept in cell biology [57], ions may flow across the membrane by active and passive mechanisms and the changes in the membrane potential trigger important physiological phenomena.

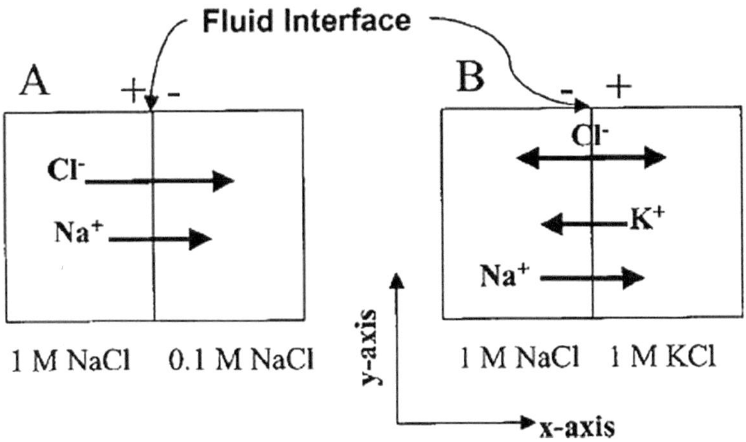

Fig. 2.3 The origin of the liquid junction potential (LJP). *Left*: chloride ions diffuse faster than sodium ions creating a potential gradient (negative at right) *Right*: there is no net flux of chloride ions but sodium diffuses faster than potassium, producing positive potential at right. Reprinted with permission from [53]

Liquid junction and membrane potential are components of the "spontaneous potential" that appears during drilling operations in the oil industry. It was one of the first logging measurements ever made and it was discovered by accident, since it caused perturbations on the electric logging systems. It records the naturally occurring voltage produced by the interaction of connate water, drilling fluid, and shale. Its usefulness was soon realized, and it remains as a useful logging measurement, after many years [58]. In this case, the "membrane" is a layer of shale or sand.

2.7.6 Electrostatics in Chemical Processing

Electrostatic phenomena are well known to chemical engineers involved with gas–solid–liquid flow systems [7], as they cause various types of operational problems and are a source of hazard. Electrostatic charging has been observed during spray-drying [59] and fluidized bed operations [60] where charge and hydrodynamics are mutually affected and excess accumulation of electrostatic charges has a severe impact on hydrodynamics [61].

2.7.7 Electrostatics in Soft Matter

Soft matter provides a large number of examples of systems whose properties and related processes largely depend on charge partition and segregation, this means they show large domains that are non-electroneutral. They include high polymers and elastomers, polyelectrolyte gels, food processing and storage, waste management, new tailored materials that easily acquire and store excess charge.

Even more important, electrostatic interactions are essential for biopolymer structure and they are thus at the heart of all the biotechnologies that currently have a great impact on human life [62]. Their relevance is now sufficient to attract the attention of theoreticians going deeply into systems that were previously often considered too complex for a proper scientific analysis. As a result, counter-intuitive phenomena like charge inversion of particles and polymers are now understood as induced by correlation [63].

2.8 Conclusions

The question in the title of this chapter can now be answered: electroneutrality is not to be often expected and if so, in very few places. It is not the rule and its conceptual use in most material systems neglects a large amount of sound scientific information.

The evolution of ionization techniques in mass spectrometry shows how great progress was achieved just by recognizing that electric charge separation takes place during simple phenomena under many conditions, in the laboratory and environment. This perception has been obstructed in many cases by the presumed validity of an "electroneutrality principle", even in the absence of significant experimental or theoretical support.

The examples given in this chapter show how charge partition takes place in macroscopic, microscopic and mesoscopic systems, under equilibrium or non-equilibrium conditions. Consequently, any time a material system is approached, we should ask ourselves which electrostatic patterns it contains and how these change with time.

Electrostatic patterns are all over, they are not rare and irrelevant oddities that are not even mentioned to most science and engineering students. Given their pervasiveness, electrostatic patterns and their consequences can and should be clearly mentioned by lecturers teaching many different disciplines and they should be considered by researchers and engineers, in every area of scientific endeavor.

References

1. Ziaei-Moayyed M, Goodman E et al (2000) Electrical deflection of polar liquid streams: a misunderstood demonstration. J Chem Educ 77(11):1520–1523
2. Marín AG, Hoeve W et al (2013) The microfluidic Kelvin water dropper. Lab Chip 13: 4503–4506
3. Xie Y, Boer HL et al (2014) Pressure-driven ballistic Kelvin's water dropper for energy harvesting. Lab Chip 14:4171–4177
4. Yang SA, Beach GSD et al (2009) Universal electromotive force induced by domain wall motion. Phys Rev Lett 102:067201

5. Tolman RC (1911) The electromotive force produced in solutions by centrifugal action. J Am Chem Soc 33:121–147
6. Lion KS (1959) Instrumentation in scientific research. McGraw-Hill, New York, p 229
7. Matsusaka S, Maruyama H et al (2010) Triboelectric charging of powders: a review. Chem Eng Sci 65:5781–5807
8. Park AA, Fan LS (2007) Electrostatic charging phenomenon in gas–liquid–solid flow systems. Chem Eng Sci 62:371–386
9. Eliasen H, Kristensen HG et al (1999) Electrostatic charging during a melt agglomeration process. Int J Pharm 184:85–96
10. Soares LC, Bertazzo S et al (2008) A new mechanism for the electrostatic charge build-up and dissipation in dielectrics. J Brazil Chem Soc 19(2):277–286
11. Rehm M, Conzen PF et al (2004) The Stewart model. 'Modern' approach to the interpretation of the acid-base metabolism. Anaesthetist 53:347–357
12. Hruška V, Jaroš M, Gaš B (2006) Simul 5—free dynamic simulator of electrophoresis. Electrophoresis 27:984–991
13. IUPAC (2016) Compendium of Chemical Terminology—the Gold Book. http://goldbook.iupac.org. Accessed 21 Jul 2016
14. Lewis NG, Randall M (1961) Thermodymanics. MacGraw-Hill, New York
15. Eisenberg B (2012) In: Rice SA, Dinner AR (eds) Crowded charges in ion channels. Advances in chemical physics, vol 148. Wiley, Hoboken, NJ
16. Rockwood AL (2015) Meaning and measurability of single-ion activities, the thermodynamic foundations of pH, and the Gibbs free energy for the transfer of ions between dissimilar materials. Chemphyschem 16(9):1978–1991
17. Fermi E (1956) Thermodynamics. Dover, New York, p 155
18. Pauling L (1988) General chemistry. Dover, New York, p 192
19. Kaupp M (2001) Chemical bonding of main group elements. In: Frenking G, Shaik S (eds) The chemical bond: chemical bonding across the periodic table, Chapter 1. Wiley–VCH, Weinheim, pp 15–16
20. Entner R (2007) Modeling and simulation of negative bias temperature instability. Dissertation, Technischen Universität Wien, Item 3.1.1. http://www.iue.tuwien.ac.at/phd/entner/node14.html. Accessed 21 Jun 2016
21. Akhukov MA, Fasolino A et al (2012) Dangling bonds and magnetism of grain boundaries in graphene. Phys Rev B 85(11):115407
22. Boehme C, Friedrich F et al (2005) Recombination at silicon dangling bonds. Thin Solid Films 487:132–136
23. Broqvist P, Alkauskas A et al (2008) Defect levels of dangling bonds in silicon and germanium through hybrid functionals. Phys Rev B 78(16):161305
24. Perera PN, Fega KR et al (2009) Observation of water dangling OH bonds around dissolved nonpolar groups. PNAS 106(30):12230–12234
25. Ma Y, Lu W et al (2012) Room temperature ferromagnetism in Teflon due to carbon dangling bonds. Nat Commun 3:727
26. Jin H, Dai Y et al (2009) Ferromagnetism of undoped GaN mediated by through-bond spin polarization between nitrogen dangling bonds. Appl Phys Lett 94:162505
27. Lyklema J, Overbeek JT (1961) Electrochemistry of silver iodide—capacity of the double layer at the silver iodide-water interface. J Coll Sci 16(6):595–608
28. Matuszewska B, Norde W et al (1981) Competitive adsorption of human plasma albumin and dextran on silver iodide. J Colloid Interf Sci 84(2):403–408
29. Bijsterbosch BH, Lyklema J (1968) Structural properties of silver iodide-aqueous solution interface. J Colloid Interf Sci 28(3/4):506–513
30. Larson I, Attard P (2000) Surface charge of silver iodide and several metal oxides. Are all surfaces Nernstian? J Colloid Interf Sci 227:152–163
31. Vonnegut B (1947) The nucleation of ice formation by silver iodide. J Appl Phys 18(7):593–595

32. Fraux G, Doye JPK (2014) Note: heterogeneous ice nucleation on silver-iodide-like surfaces. J Chem Phys 141:216101
33. Gobinathan R, Ramasamy P (1983) Ice nucleation at the silver iodide-aqueous solution interface. Mat Res Bull 18:593–600
34. Hirabayashi A, Sakairi M et al (1994) Sonic spray ionization method for atmospheric pressure ionization mass spectrometry. Anal Chem 66:4557–4559
35. Burgo TAL, Galembeck F (2015) On the spontaneous electric-bipolar nature of aerosols formed by mechanical disruption of liquids. Colloids Interf Sci Commun 7:7–11
36. Willians MW (2011) Triboelectric charging of insulators—evidence for electrons versus ions. IEEE Trans Ind Appl 47(3):1093–1099
37. Eliasen H (1999) Kristensen process. Int J Pharmaceut 184:85–96
38. Wang J, Zhao J et al (2015) Lupine protein enrichment by milling and electrostatic separation. Innov Food Sci Emerg 33:596–602
39. Bailey AG (1993) Charging of solids and powders. J Electrostat 30:167–180
40. Wang J, De Wit M et al (2014) Analysis of electrostatic powder charging for fractionation of foods. Innov Food Sci Emerg 26:360–365
41. Heinicke G (1984) Tribochemistry. Carl Hanser Verlag, München–Wien
42. Baláž P (2008) Mechanochemistry in nanoscience and minerals engineering. Springer, Berlin
43. Beyer MK, Clausen-Schaumann H (2005) Mechanochemistry: the mechanical activation of covalent bonds. Chem Rev 105(8):2921–2948
44. Fernández-Bertran JF (1999) Mechanochemistry: an overview. Pure Appl Chem 71(4):581–586
45. Gilman JJ (1996) Mechanochemistry. Science 274:65
46. Boldyreva E (2013) Mechanochemistry of inorganic and organic systems: what is similar, what is different? Chem Soc Rev 42:7719–7738
47. James SL, Adams CJ et al (2012) Mechanochemistry: opportunities for new and cleaner synthesis. Chem Soc Rev 41:413–447
48. Friščić T (2012) Supramolecular concepts and new techniques in mechanochemistry: cocrystals, cages, rotaxanes, open metal–organic frameworks. Chem Soc Rev 41:3493–3510
49. Craig SL (2012) Mechanochemistry: a tour of force. Nature 487:176–177
50. Ribas-Arino J, Shiga M et al (2009) Understanding covalent mechanochemistry. Angew Chem Int Ed 48:4190–4193
51. Barry PH, Lynch JW (1991) Liquid junction potentials and small cell effects in patch-clamp analysis. J Membr Biol 121:101–117
52. Dryfe RAW (2007) Liquid junction potentials. In: Zoski C (ed) Handbook of electrochemistry, Chapter 20. Elsevier, Amsterdam, pp 849–877
53. Bunakova LV, Khanova LA et al (2004) Water-solvent liquid junction potential for some low-dielectric solvents. J New Mat Electrochem Syst 7(3):241–245
54. Munson MS, Cabrera CR et al (2002) Passive electrophoresis in microchannels using liquid junction potentials. Electrophoresis 23:2642–2652
55. Graham DJ, Jaselskis B et al (2013) Development of the glass electrode and the pH response. J Chem Educ 90:345–351
56. Rover Junior L, Garcia CAB et al (1998) Acetylsalicylic acid determination in pharmaceutical samples by FIA-potentiometry using a salicylate-sensitive tubular electrode with an ethylene-vinyl acetate membrane. Anal Chim Acta 366:103–109
57. Wright SH (2004) Generation of resting membrane potential. Adv Physiol Educ 28:139–142
58. Glover PWJ (2016) Petrophysics MSc Course. Notes: The spontaneous potential log, chapter 18, p 218. Department of Geology and Petroleum Geology, University of Aberdeen, UK. https://groups.google.com/forum/#!msg/msc13iitkgp/NgvmuYtZm6A/ESXFV8LfSAMJ. Accessed 20 Sept 2016
59. Murtomaa M, Savalainen M, Christiansen L, Rantanen J, Laine E, Yliruusi J (2004) Static electrification of powders during spray drying. J Electrostat 62(3):63–72
60. Mehrani P, Bi HT et al (2005) Electrostatic charge generation in gas-solid fluidized beds. J Electrostat 63:165–173

61. Dong K, Zhang Q, Huang Z, Liao Z, Wang J, Yang Y (2015) Experimental investigation of electrostatic effect on bubble behaviors in gas-solid fluidized bed. AIChE J 61(4):1160–1171
62. Holm C, Kékicheff P et al (2000) Preface in electrostatic effects in soft matter and biophysics. NATO Science Series II, vol 46. Kluwer, Dordrecht, p v
63. Nguyen TT, Yu A et al (2000) Lateral correlations of multivalent counterions is the universal mechanism of charge inversion. In: Holm C, Kékicheff P et al (eds) Preface in electrostatic effects in soft matter and biophysics. NATO Science Series II, vol 46. Kluwer, Dordrecht, pp 469–485

Chapter 3
Charge Carriers Within the Atomic-Molecular Theory

Contents

3.1 Charge and Matter

Electric charge is a conserved fundamental property of matter. It can be positive or negative. It is found in discrete quantities accounting for the forces applied on matter by electromagnetic fields, following the standard current view.

The dictionary definition above is satisfactory for all the purposes of this book and it will be used throughout. However, the authors acknowledge that it leaves open questions that have stimulated further thoughts, e.g. concerning the nature of charge [1]. This leads to a broader discussion on the nature of science that goes well beyond the scope of this book. A fact that is especially intriguing is the following:

F. Galembeck, T.A.L. Burgo, *Chemical Electrostatics*,
DOI 10.1007/978-3-319-52374-3_3

even though charge on ordinary matter is always a positive or negative integer multiplied by the charge of the electron *e*, the charge of quarks may be equal to 2/3 or −1/3 times *e*, this means, some sub-atomic particle-energy packages have fractional charge [2].

The conceptual framework of this chapter is the atomic-molecular theory that has proven adequate for the study of matter as it exists in the environments where life thrives and humans live. This means, the simplest charged entities considered here are protons, electrons, neutrons, and ions.

3.1.1 Protons, Electrons, Neutrons, Molecules, and Ions

Matter found on Earth surface is formed by atoms of the elements represented in the Periodic Table, most often bound together to form molecules, ions and metals, packed in crystals of various types or moving freely in liquids and gases. These are in turn formed by protons, neutrons, and electrons that usually preserve the atomic arrangements but may be at least partly dismantled when high-energy particles impinge on them or when they acquire sufficiently high energy to form a plasma.

The energy from radiation absorbed by matter may be sufficient to eject one or more electrons, leaving behind positive ions or excess positive charge in crystals. This is the photoelectric effect that takes many forms, from the photovoltaic cells for energy production to the X-ray photoelectron analytical spectrometers widely used in research and analysis and still, to the formation of the ionosphere.

On the other hand, the affinity of electrons for various elements is highly variable. Coarsely, it increases from the left to the right hand side of the Periodic Table, so that halogens, chalcogens, and other non-metals are able to keep excess electrons forming negative ions. Electronegativity decreases from the top to the bottom of the Table, enough to allow fluorine atoms to establish bonds with the larger noble gases, like xenon.

There are also many stable polyatomic ions like sulfate and macromolecular ions represented by diverse natural substances as DNA and pectin or synthetic ones such as polyacrylate and polysulfonate. These are very stable species but many others are much less so, like the reactive propagating chains in anionic and cationic polymerization.

3.1.2 Formation and Stability of Ionic Species

Ionic species abound on Earth and they are widespread in the oceans and rocks that make most of the crust, as well as the biosphere. On the other hand, they are major components of the ionosphere and they also play an important role in the moist troposphere that was recently shown to act as a charge reservoir [3].

The simplest case of formation of ionic species is the breakdown of a polar covalent bond, forming separate cations and anions. This happens with simple substances like the carbonic acid from the atmosphere and the most complex nucleic acids, soluble molecules like acetic acid and insoluble ones as silicates.

Another simple way to form ionic species is a redox reaction, when different neutral atoms concurrently loose and acquire electrons, forming positive and negative ions. This is observed during metal corrosion, when the metal loses electrons to adjacent oxygen molecules or water, forming oxide (O^{2-}) or hydroxide (OH^-) ions. In some cases of metal corrosion, the released electron is captured by H^+ ions, forming molecular hydrogen. Oxygen is a prolific source of negative ions: beyond oxide and hydroxide there are also peroxide (O_2^{2-}) and superoxide (O_2^-) ions. Moreover, oxygen participates from most covalent bonds that undergo ionization, as mentioned in the previous paragraph.

Electron supply and withdrawal forming ions or transforming them back into neutral species can also take place from some external electric power supply connected to an electrochemical cell. This is observed in the cells and batteries that power everyday life and in many important electrochemical process industries that make chlorine and soda, aluminum, copper and many other metals, hydrogen peroxide and a host of other products.

3.1.2.1 Ions from Water

Throughout this book, the ionization of water will receive great attention. The division of water molecules into hydronium (H^+) and hydroxide (OH^-) ions takes place to a very limited extent: under ambient conditions, only one out of 10^7 molecules ionize, one-tenth of a part per million. However, given the magnitude of Avogadro number, this amounts to ca. 10^{17} positive and negative ions per mol of ultrapure, neutral water, only 18 g.

This is not sufficient to allow water to be classified as a conductor of electricity, for most purposes. On the other hand, if the positive and negative ions in 1 mol of water were somehow separated, their charge would amount to 10^{-2} C each, given the Faraday constant, 96,485 C/mol. This amount of charge, flowing for a short time as for instance 1 ms produces 10 A current. Consequently, there is much electricity in a few grams of water, one of the most abundant substances in the Earth crust.

3.1.3 Electrons

Electrons are the prevalent charge carriers in metals and semiconductors. They are the charge carriers in n-type semiconductors while in p-type semiconductors the "holes" are actually electron vacancies whose motion is ultimately a way to describe the motion of electrons themselves.

In solid or liquid metals, the outer electrons move freely across the setup of cations. The latter are also displaced in an electric field but making a much lower contribution to the electric current [4] than electrons. In pure semiconductors, free electrons are present in lesser amounts. They also play an important role in ionized gases and in any plasma.

Electrons from heated lamp or electron emitter filaments leave the metal surface toward the gas phase. This is thermionic emission, usually done under vacuum to prevent filament burning and also due to the intense electron scattering in the atmosphere. Another way to produce an electron current is field emission from a sharp tip of a metal, kept under a few kilovolts negative potential. The resulting current is sufficient for image formation in microscopy, displays and even for electron-beam lithography.

On the other hand, free electrons are short-lived in many material environments and they do not keep their identity for long in most gases, liquids, and solids. This is familiar to workers in many research areas, like electron microscopy. Handling electrons in gas phase is only possible under very low pressures, since their mean free-paths are very short, considering both elastic and inelastic scattering. Moreover, samples for transmission electron microscopy are necessarily thin, often well below 100 nm, while 1 μm samples are useful but in a few cases only. This shows that electrons can hardly contribute to space charge in condensed phase dielectrics: once they enter an insulating solid or liquid, they will quickly undergo multiple scattering, losing energy [5] until this reaches a value lower than the electron affinity of some sample component.

3.2 Charge Motion

Charge motion in different media covers a broad range of velocities, since it is always coupled to motion of carriers that have finite mass. Consequently, electrical resistivity that depends largely on charge motion covers 20 orders of magnitude [6].

Charges on insulators are usually immobile, a statement that is amply verified by the large number of published Kelvin micrographs and electric potential maps acquired using Kelvin electrodes, as for instance in the examples given in Figs. 3.1 and 3.2 [7]. Even the Kelvin micrographs obtained with the first generation of commercial instruments showed steady potential patterns with 10-nm spatial resolution whose acquisition was repeated over and over in a process taking up to a few hours. Image stability was only possible if the speed of charge displacement was less than 10 nm per hour or so, corresponding to less than 10 pm per second. This may be compared to electron thermal velocities in Si and Ag that are respectively 2.3×10^5 m s^{-1} and 16.7×10^7 m s^{-1}, 14 and 19 orders of magnitude higher. On the other hand, diffusion coefficients for common ions in water are seven orders of magnitude lower than electrons in metals, as shown in Table 3.1.

Electrostatics is largely concerned with the systems where charge is immobile or it moves slowly. However, since charges do not move instantly even in highly conducting metals, the onset of electrostatic phenomena depends on the timescale of the observations. For instance, a classical estimate of electron displacement in a metal surface probed with visible light photons may extend to 1 nm only, during one period of the light wave, or else, bulk metal is not electrically disturbed by the incidence of light.

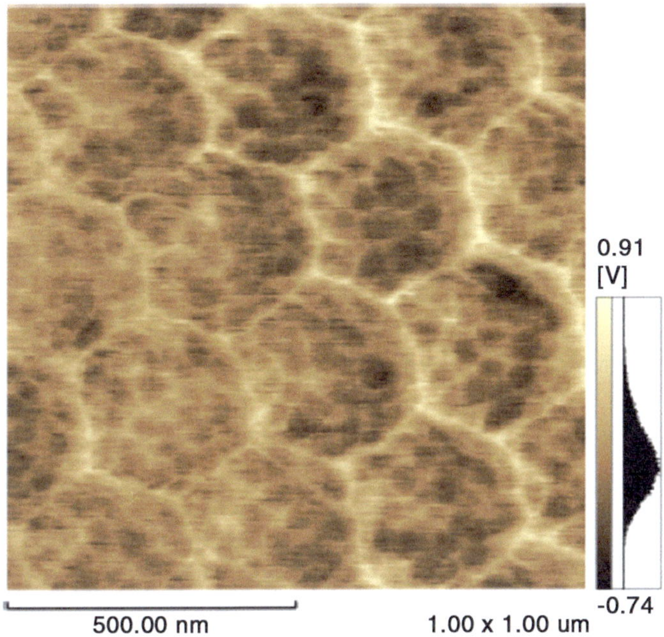

Fig. 3.1 Kelvin micrograph of poly(styrene-co-hydroxyethylmethacrylate) (PS-HEMA) dry latex. Dark (*bright*) pixels contain excess negative (*positive*) charge. The pixel electric potential histogram is at the right. Reprinted with permission from [3]

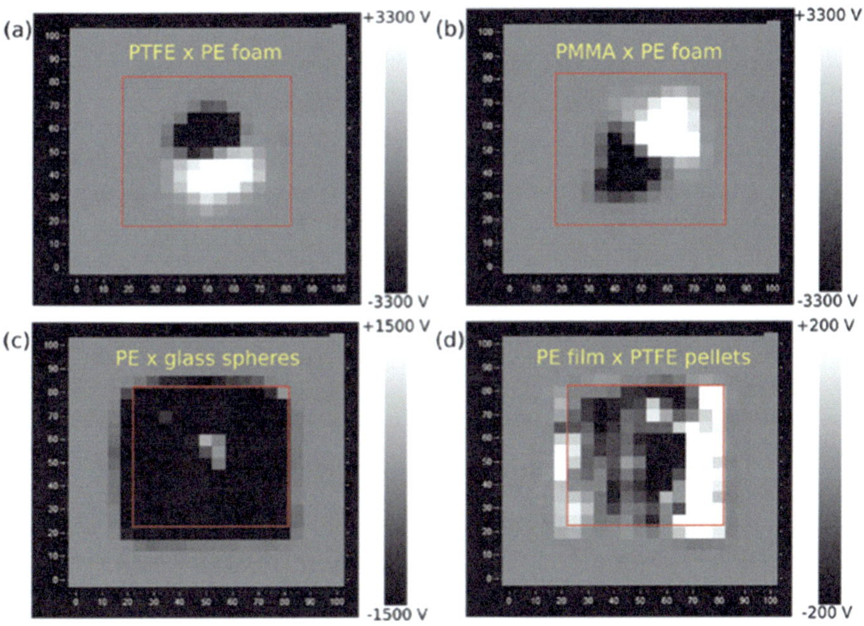

Fig. 3.2 Macroscopic potential maps from tribocharged samples: (**a**) polytetrafluoroethylene (PTFE) and (**b**) poly(methyl methacrylate) (PMMA) rubbed with polyethylene (PE) foam disks. PE film abraded with (**c**) glass spheres and (**d**) PTFE pellets in a planetary mixer. *Red squares* indicate the polymer area on each image and the *x*, *y* axes are in millimeters. Positive and negative domains covering many millimeters or centimeters are observed in every case. Reprinted with permission from [7]

Table 3.1 Mobility data for electrons and holes in Si, electrons in Ag and Na^+ ions in aqueous solution at infinite dilution, 25 °C

Property	Si, electrons [8]	Ag, electrons [9]	Na^+, ions
Breakdown field V/cm	$\approx 3 \times 10^5$		
Mobility ($cm^2\ V^{-1}\ s^{-1}$)	≤ 1400	9490	5×10^{-8}
Mobility holes ($cm^2\ V^{-1}\ s^{-1}$)	≤ 450		
Diffusion coefficient ($cm^2\ s^{-1}$)	≤ 36	241	13.3×10^{-6}
Diffusion coefficient holes ($cm^2\ s^{-1}$)	≤ 12		
Electron thermal velocity ($m\ s^{-1}$)	2.3×10^5	16.7×10^7	
Hole thermal velocity ($m\ s^{-1}$)	1.65×10^5		

3.2.1 Water

Accounts of the electrical properties of water usually emphasize its large dielectric constant and the large dipole moment that is often related to the concerted orientation of the large dipole moments of water molecules. In other context, there is a persistent discussion on the structure of liquid water that was initiated by efforts to interpret the Hofmeister effect. This is the collective name for the effects of dissolved ions in various types of experiments, whose models and mechanisms have been discussed for decades, with no signs for settling down [10].

Another complex situation is found in the case of the models and mechanisms put forward to explain the abnormally large contribution made by H^+ and OH^- ions to water electrical conductivity, when the Grothuss mechanism is often invoked [11, 12] but it has also been challenged [13]. Following Marcus [14], the enhanced conductivity behavior of H^+ and OH^- ions is sustained in aqueous co-solvent mixtures under certain conditions. For instance in aqueous acetonitrile, proton hopping is observed only above 20% water, when clusters of a minimal size are present [15].

Summing up, water plays a unique role in electrostatics, for various reasons: its conductivity is low but significant and it is greatly increased by dissolving ionic solids. Moreover, it offers unique possibilities for charge migration, thanks to the Grothuss mechanism or its alternative models. Most important, it is by itself an important reservoir of H^+ and OH^- ions.

3.2.2 Ionic Liquids

Ionic liquids have been known for decades but they only attracted great attention recently. For this reason, it is not surprising that most data on their basic properties including ion mobility have been obtained recently.

A rich collection of data is in [16], showing the specific conductivity of a number of representative ionic liquids. The room temperature conductivity, σ, is always within a broad range of 0.1–18 mS/cm. Conductivity of the order of 10 mS/cm is typical of ionic liquids based on $[EtMeIm]^+$, while ionic liquids based on tetraalkylammonium, pyrrolidinium, piperidinium, and pyridinium cations are characterized by lower conductivity, in the 0.1–5 mS/cm range. Thus, the highest room temperature ionic liquid conductivity is much lower than common aqueous electrolyte solutions. For example, the specific conductivity of aqueous KOH (29.4 wt.%) solution is 540 mS/cm and in the electrolyte in lead-acid batteries, 30 wt.% aqueous H_2SO_4, it is ca. 730 mS/cm. Non-aqueous ionic solutions show one order of magnitude lower conductivity, for instance $\sigma = 60$ mS/cm for Et_4NBF_4 in acetonitrile. Dilution may produce an increase in conductivity, and in some cases the conductivity initially increases with the increasing amount of the salt, goes through a maximum and then decreases. All these changes are strongly dependent on the viscosity.

A study on *N*-butyl-*N*-methylpyrrolidinium cation with bis(trifluoromethanesulfonyl) imide, bis(pentafluoroethanesulfonyl) imide and (trifluoromethanesulfonyl) (nonafluorobutanesulfonyl) imide provided data on the pure ionic liquids and their mixtures. Viscosity at 303 K is in the 60–300 mPa s range while the diffusion coefficients range is 5–25 × 10^{-12} m^2 s^{-1}, few orders of magnitude lower than the diffusion coefficients for simple ions in water [17].

Following the Nernst–Einstein equation (Eq. 3.1), the ionic conductivity in these liquids is also much lower than in solutions of simple ions.

$$= z^2 e_0 FD / k_B T = z^2 N_A e^2_0 D / k_B T \qquad (3.1)$$

Various aspects of the thermodynamics of non-aqueous mixtures containing ionic liquids were reviewed in 2005, showing interesting data on activity coefficients. This work also revealed considerable lack of data on transport properties such as viscosity, diffusion coefficients, heat, and electrical conductivity in ionic liquids + solvent mixtures. Data of other properties like surface and interphase tension were then completely missing but they are needed for a fuller understanding of ion transport properties [18]. Earlier, equivalent conductance of ionic liquids in non-aqueous liquids and the limiting ion conductance $\lambda^{\circ}_{\mathrm{i}}$ were found in quantitative agreement (within ±2%) with the values predicted by an empirical equation [19].

3.3 Charge Carriers at Interfaces

Interfaces are always important places for the appearance of excess charge, due to various mechanisms that will be discussed in Chap. 5. In this section, we examine some issues related to charge motion and storage at or across interfaces.

3.3.1 Dimensionality

The geometric dimension of smooth interfaces like most liquid–liquid or liquid–air interfaces is *2* while most solid–liquid or solid–gas interfaces are fractal, with dimension in the range between *2* and *3*. For this reason, diffusion and thus charge mobility at interfaces may be significantly different than in the bulk phase.

There is not sufficient data in the literature to allow broad statements to be made but interesting effects have been reported. For instance, Brownian dynamics simulations of colloidal hard spheres in 2- and 3D showed smaller diffusion coefficients of the $d = 2$ samples at small concentrations, whereas the opposite is found at very high concentrations above *volume fraction* = *0.55*. The absence of the third degree of freedom of motion in $d = 2$ leads to more effective caging of the particles while in $d = 3$ the in-plane cage can relax also by particle motion in the z direction. The contribution of charge to this problem is probably very large but there is not currently information to discuss this.

3.3.2 Electrodes and Electrochemistry

Charge carriers within a bulk phase are of one or two types: electrons and ions in metals and gases, electrons and holes in intrinsic semiconductors but either electrons or holes in n- or p-type semiconductors, respectively. In electrolyte solutions, molten salts and ionic liquids carriers are positive and negative ions.

So, when electric current flows across different phase types, conversion from one to another type of carrier is needed. This is typically the case in electrolyte cells, piles, and batteries. A simple and important case is copper metal refining by electrolysis, in cells consisting of two copper electrodes and a copper salt solution. Under a potential difference applied to the two electrodes, electrons are withdrawn from the positive electrode that releases the resulting copper (II) ions. These in turn migrate through the solution toward the negative electrode, where they are deposited leaving behind impurities that settle in the bottom of the container.

In this case, current flow is assured by the two electrode reactions:

$$\text{at the positive electrode: } Cu = Cu^{2+} + 2\,e^{-}$$
$$\text{at the negative electrode: } Cu^{2+} + 2\,e^{-} = Cu$$

However, chemical kinetics often makes the picture more complex. A noted case is that of H^+ ion discharge in a negative electrode: it is a fast reaction at a platinum electrode potential slightly more negative than the equilibrium potential but it is very slow when platinum is replaced by a mercury electrode, when an overpotential as high as −0.78 V is required to achieve deposition of hydrogen, a fact that was once important in the chlorine—soda electrochemical cells: when a brine is electrolyzed using a mercury cathode, a sodium amalgam is formed since Na^+ ions are discharged prior to H^+ ions [20], at least when sodium concentration in the amalgam is not too high. When a graphite anode is used in the same process, chlorine and/or oxygen evolve, depending on the relative concentrations of Cl^- and OH^- ions in the brine. Currently, efforts to understand and to lower overpotentials are actively pursued in many research contexts, often connected to water oxidation and other reactions relevant to solar energy harvesting and sustainable energy production [21].

These examples illustrate the complexity of electric current flow across a circuit where both electrons and ions are serial current carriers. This is the focus of the attention of many electrochemists, a group of scientists and engineers that are continuously making huge contributions to science and technology.

3.3.3 Electrodes in Capacitors

Electrodes store charge in capacitors and the amount of stored charge depends on the dielectric polarization, the applied voltage and the geometry of the electrode set-up. For metal electrodes separated by a simple and uniform dielectric, charge carriers are electrons fed to one electrode and withdrawn from the other.

However, in electrolytic and supercapacitors, the energy stored is used to separate ions of opposite signal, kept adjacent to the two electrodes. This topic will be dealt further in Chap. 4, on charges at interfaces.

3.3.4 Ion-Exchange Membranes

These are films or tubing made with a polymer that carries covalently bound anions like carboxylate, phosphate or sulfonate, or cations like quaternary alkylammonium. In a cation-exchanging membrane, anions are fixed but cations can cross the membrane when driven by a potential or concentration gradient.

For this reason, these membranes contribute their selectivity to transform the usually bidirectional flow across an electrolyte solution into a unidirectional flow [22]. This is invaluable in chemical separation, seawater desalination, and wastewater treatment.

3.3.5 Gas–Liquid and Gas–Solid Interfaces

The presence of ions in the atmosphere is frequently neglected and it is not rare to see statements on charge in gas–liquid interfaces as solely dependent on excess concentration of solute species, following Langmuir adsorption isotherm. Excess charge is pronounced in the surfaces of surfactant solutions and monolayers, when anionic (cationic) surfactants impart negative (positive) charge to the interface. Other important gas interfaces are those of hydrogen electrodes and similar devices, where electronic charge is converted to ionic charge and vice versa.

Gas–solid interfaces in dielectrics and in grounded conducting or semi-conducting solids are often considered to be devoid of excess charge. In biased metals they have positive or negative excess of electrons, depending on the applied potential. Surface charge excess also appears due to induction, in conducting solids.

Beyond these well-established situations, there are many others that are only partially understood. These are presented in Chaps. 6, 7, and 9.

3.3.5.1 Vapor Electricity

In 1840, an unusual electrical shock incident few miles from Newcastle (UK) led Lord Armstrong to send a description of this unprecedented electrical phenomenon to Faraday. Shortly, an engine-man, working on a faulty boiler, placed his hand in a leaking steam while his other hand was on a metal valve and this caused a violent electrical discharge [23]. Using an electrometer, Faraday found the engine steam to be positively charged and later he proposed that friction of steam against other materials was responsible for charge partition between vapor and the contacting materials [24]. This experiment is more often mentioned than reproduced and it has no satisfactory explanation, yet. Faraday's suggestion is still intriguing but it can hardly be discussed because "friction", by itself, is not yet a developed topic within the atomic-molecular theory framework.

Later in the nineteenth century, Lord Kelvin created the well-known *Kelvin's thunderstorm* [25]. Most reports on these often-mentioned experiments are just qualitative but they have recently reached a high degree of sophistication in microfluidic devices [26]. This phenomenon is probably related to serious problems observed while handling liquid fuels [27]: large crude carriers sank or suffered severe damage following explosions caused by sparks produced during tank washing with steam [28]. This is just one case of still frequent, serious accidents that are ultimately caused by the absence of clear understanding of the involved basic events (see Chap. 12).

3.4 Ions, Electrons, or Both?

The nature of charge carriers in dielectrics has been debated in the literature [29] for decades and some authors have clearly stated the lack of consensus on that matter [30–32]. Important results showing the participation of ions have been published by the Whitesides group [33, 34] and by others, including the authors' group. At about the same time, other results by the Bard group [35, 36] pointed toward charging by electrons. The situation was summed up in 2012 by Williams [37], who considered also a third mechanism that is mass transfer. This will be further discussed in Chaps. 6 and 9 in this book.

The important area of electrets provides a wealth of information on the techniques for charge trapping and migration within solids, including profiles for charge distribution but the nature of charge traps receives only minor attention. The participation of both ions and electrons is acknowledged [38, 39] but many authors refer to "space charge" without referring to the nature of charge bearing species. An important contribution was the development of micro-foam electrets [40, 41], based on the formation of ordered arrays of charge dipoles within cellular polymers. These are formed by poling microfoams and thus ionizing the cell contents, followed by the deposition of charged species on opposite walls of the same cell. This is a possibility to obtain polymer film domains with positive or negative charge separately, allowing the identification of charge-bearing species.

This discussion became more complex following two discoveries. First, triboelectrified insulator surfaces display complex charge and potential patterns showing the coexistence of domains with opposite charge, side by side. Moreover, experimental evidence revealed that mass transfer is often associated with contact and tribocharging. This will be further discussed in Chap. 8.

At this point in time, it looks like different mechanisms operate in different systems and perhaps two different mechanisms may work simultaneously, in some cases.

References

1. Krasnoholovets V (2003) On the nature of the electric charge. Hadronic J Suppl 18:425–456
2. Novaes SF (2000) Standard model: an introduction. http://arxiv.org/abs/hep-ph/0001283. Accessed 23 Jun 2016
3. Rezende CA, Gouveia RF et al (2009) Detection of charge distribution in insulator surfaces. J Phys Condens Matter 21(16):263002
4. Verhoeven J (1963) Electrotransport in metals. Metall Rev 8(1):311–368
5. Werner WSM (2001) Electron transport in solids for quantitative surface analysis. Surf Interf Anal 31:141–176
6. Heaney MB (2004) Electrical conductivity and resistivity. In: Webster J (ed) Electrical measurement, signal processing, and displays, chapter 7. CRC Press LLC, Boca Raton, FL
7. Burgo TAL, Ducati TRD et al (2012) Triboelectricity: macroscopic charge patterns formed by self-arraying ions on polymer surfaces. Langmuir 28:7407–7416
8. Si—Silicon electrical properties. http://www.ioffe.ru/SVA/NSM/Semicond/Si/electric.html. Accessed 3 Jun 2016
9. Palenskis V (2013) Drift mobility, diffusion coefficient of randomly moving charge carriers in metals and other materials with degenerated electron gas. World J Condens Matter Phys 3:73–81
10. Salis A, Ninham BW (2014) Models and mechanisms of Hofmeister effects in electrolyte solutions, and colloid and protein systems revisited. Chem Soc Rev 43:7358–7377
11. Cukierman S (2006) Et tu, Grotthuss! and other unfinished stories. Biochim Biophys Acta 1757:876–885
12. Krikštopaitis JA (2006) Concerning the origins of charge transfer in the micro-structure of matter: the contribution of Theodor von Grotthuss. Electrochim Acta 51:5999–6002
13. Agmon N (1995) The Grotthuss mechanism. Chem Phys Lett 244:456–462
14. Marcus Y (2015) Ions in solution and their solvation. Wiley, Hoboken, NJ
15. Gileadi E, Kirowa-Eisner E (2006) Electrolytic conductivity—the hopping mechanism of the proton and beyond. Electrochim Acta 51:6003–6011
16. Galiński M, Lewandowski A et al (2006) Ionic liquids as electrolytes. Electrochim Acta 51:5567–5580
17. Castiglione F, Raos G et al (2010) Blending ionic liquids: how physico-chemical properties change. Phys Chem Chem Phys 12:1784–1792
18. Heintz A (2005) Recent developments in thermodynamics and thermophysics of non-aqueous mixtures containing ionic liquids. A review. J Chem Thermodyn 37:525–535
19. Gill DS, Sekhri MB (1992) New approach to the evaluation of single-ion conductances in pure and mixed non-aqueous solvents. [Part 2]. J Chem Soc Farad Trans 1(78):119–125
20. Moody B (1991) Comparative inorganic chemistry. Arnold, London, p 127
21. Zhang T, Wang C et al (2014) biomimetic copper water oxidation catalyst with low overpotential. J Am Chem Soc 136:273–281
22. Tanaka Y (2015) Ion exchange membranes: fundamentals and applications. Elsevier, Amsterdam, p 47
23. Armstrong WG (1840) On the electricity of a get of steam issuing from a boiler. Philos Mag Ser 3 17(111):370–374
24. Faraday M (1840) Experimental researches in electricity. Eighteenth series. Philos Trans R Soc Lond 133:17–32
25. Thomson W (1867) On a self-acting apparatus for multiplying and maintaining electric charges, with applications to illustrate the voltaic theory. Proc R Soc Lond 16:67–72
26. Marín ÁG, van Hoeve W et al (2013) The microfluidic Kelvin water dropper. Lab Chip 13:4503–4506
27. Finke J (1989) Electrostatic effects of charged steam jets. J Electrostat 23:69–78
28. Jones MRO, Bond J (1984) Electrostatic hazards associated with marine chemical tanker operations—criteria of inflammability in tank cleaning operations. Chem Eng Res Des 62:327–333
29. Loeb LB (1945) The basic mechanisms of static electrification. Science 102(2658):573–576

30. Schein LB (2007) Recent progress and continuing puzzles in electrostatics. Science 316(5831): 1572–1573
31. Castle GSP (1987) Contact charging between insulators. J Electrostat 40, 41:13–20
32. Bailey AG (2001) The charging of insulator surfaces. J Electrostat 51, 52:8290
33. McCarty LS, Whitesides GM (2008) Electrostatic charging due to separation of ions at interfaces: contact electrification of ionic electrets. Angew Chem Int Ed 47:2188–2207
34. McCarty LS, Winkleman A et al (2007) Ionic electrets: electrostatic charging of surfaces by transferring mobile ions upon contact. J Am Chem Soc 129:4075–4088
35. Liu CY, Bard AJ (2008) Electrostatic electrochemistry at insulators. Nat Mater 7:505–509
36. Liu CY, Bard AJ (2009) Electrons on dielectrics and contact electrification. Chem Phys Lett 48:145–156
37. Williams MW (2012) What creates static electricity? Am Sci 100(4):316–325
38. Kressmann R, Sessler GM, Gunther P (1996) Space-charge electrets. IEEE Trans Dielect Elec Insul 3(5):607–623
39. Bauer-Gogonea S, Bauer S (2001) Polymer electrets for electronics, sensors, and photonics. In: Nalwa HS (ed) Handbook of advanced electronic and photonic materials and devices, vol 10. Academic, New York, pp 185–224
40. Gerhard-Multhaupt R (2002) Voided polymer electrets—new materials, new challenges, new chances. In: 2002 11th international symposium on Electrets (ISE11) Proceedings, pp 36–45
41. Hsu TH, Yeh CN, Su YC (2010) Piezoelectric PDMS electrets for MEMS transducers. In: 2010 IEEE 23rd international conference on Micro Electro Mechanical Systems (MEMS), pp 388–391. http://www.ieee.org/conferences_events/conferences. Accessed 23 Jun 2016

Chapter 4
Charge at Interfaces

Contents

4.1 The Maxwell-Wagner-Sillars Effect

Charge accumulation can take place at any interface, following the Maxwell-Wagner-Sillars effect that is observed whenever current flows across the interface of two materials [1].

Two parameters, dielectric constant ε and conductivity σ, define the macroscopic electrical properties of materials and their ratio $\tau = \varepsilon/\sigma$ is the relaxation time for each equivalent electric circuit given by the parallel capacitance and resistance of each material. When an electric current j crosses the interface, the difference in the relaxation times produces charge accumulation at the interface. This is a macroscopic dielectric polarization process that is called interfacial polarization.

F. Galembeck, T. A. L. Burgo, *Chemical Electrostatics*,
DOI 10.1007/978-3-319-52374-3_4

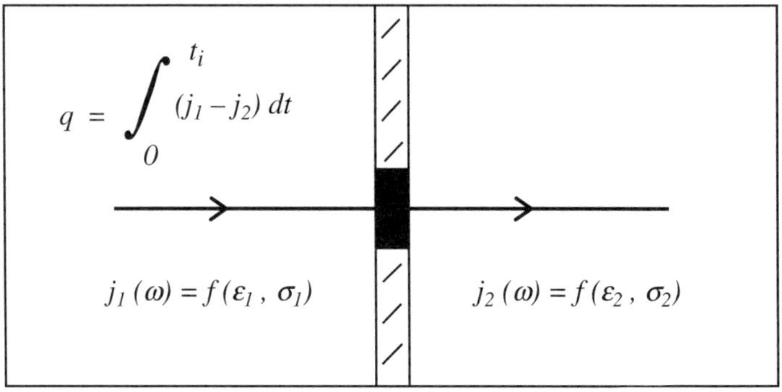

Fig. 4.1 Schematic description of the Maxwell-Wagner-Sillars effect

The time constant for charge accumulation is different from the time constants for each contacting material forming a series circuit, because it also depends on geometric factors. This is understood considering that the contribution of each material to the capacitance and impedance of the series circuit depends not only on the respective values for ε and σ but also on film thickness and interfacial area. This is represented in Fig. 4.1.

Given the non-specificity of the Maxwell-Wagner-Sillars effect, it should be expected to take place in every kind of interface, thus producing interfacial polarization in any interface.

4.2 Solid–Liquid Interfaces

Solid–liquid interfaces acquire charge for many additional reasons: ionization of solid surface groups, selective adsorption/desorption of ions from the solid surface, and differential solubility of ions from an ionic compound. Selective adsorption of H^+ or OH^- ions from water has paramount importance, because is found in most natural environments.

4.2.1 Mechanisms for S/L Interface Charging

Ionization of solid surface groups is expected in acidic metal oxides and in polymers containing carboxylic groups, like cross-linked poly(styrene sulfonate) resins. On the other hand, polyethylene surfaces are more or less oxidized [2] and they may contain pendant carboxylic groups that undergo ionization in the presence of water and other solvents. When these systems are considered, it is important to keep in mind that the pK_a for the -COOH dissociation reaction is not a constant, as in low-MW solutes. Instead, pK_a for ionizable groups bound to a polymer chain or a surface varies with the degree of ionization [3], due to the effect of the increased negative charge density on the retention of H^+ ions leaving the surface.

Differential solubility is observed in salts and it is often neglected because chemistry teachers normally introduce the topics of solubility and the solubility product K_{ps} using an oversimplified picture. A frequent example is silver iodide and its solubility equilibrium is represented by Eq. (4.1).

$$AgI = Ag^{+} + I^{-} \text{ with } K_{ps} = \left[Ag^{+}\right]\left[I^{-}\right] = 8.52 \times 10^{-17} \quad (4.1)$$

This suggests that the remaining undissolved AgI particles are electroneutral and the saturated solution contains equal concentrations of Ag^+ and I^- ions, yielding $p_{Ag} \approx 8$. Indeed, AgI particles are only neutral when $p_{Ag} = 5.5$ and $p_I = 10.5$, this means, the two concentrations differ by five orders of magnitude and both concentrations depart from equilibrium by nearly three orders of magnitude.

This is easily understood, considering that unsolvated silver cations are significantly smaller (129 pm radius) [4] and thus more solvated [5] by water than iodide (206 pm). Consequently, silver ion activity is lower within water than the activity of iodide, at the same concentration and Ag^+ actually exhibits extreme solvation effects that are not observed for any other ions except Zn^{2+} and Au^+ [6].

Selective adsorption of ions on the solid surface is easily observed with surfactant [7] and polyelectrolyte solutions. For instance, immersion of most solids on aqueous solutions of anionic surfactants like sodium lauryl sulfate leads to strong adsorption of the lauryl sulfate anions, much more intensely than Na^+. This imparts to the surface a large negative charge that is one of the main factors of the dispersant capabilities of this and other surfactants, including detergency. The hydrophobic interactions responsible for surfactant adsorption largely overcome the electrostatic repulsion among adjacent adsorbed ions allowing the formation of adsorbed multilayers, hemimicelles [8, 9], and solloids[1] [10]. For these reasons, surfactants and their mixtures can drastically change the interfacial properties and hence they are used in many industrial processes such as dispersion/flocculation, flotation, emulsification, corrosion inhibition, cosmetics, drug delivery, chemical mechanical polishing, enhanced oil recovery, and nanolithography. On the other hand, since surface-active substances are widespread in any terrestrial environment and most are ionic compounds, they certainly play an important role in imparting charge to S/L interfaces.

4.2.2 *The Electric Double Layer*

Thus, a charged solid surface immersed in a liquid usually acquires excess charge and this interferes with the distribution of ions in the liquid forming an *electrical double layer* [11, 12], a peculiar, complex environment (Fig. 4.2). The water molecule dipoles adjacent to the surface are oriented and ions with opposite charge may bind strongly to the surface forming the *Stern layer* while other ions located beyond the *Stern plane* form the diffuse part of the double layer (or *Gouy-Chapman layer*), that obeys the Poisson-Boltzmann distribution of charge density. Strongly bound solvent

[1] Solloids are surface-mediated colloids or colloids made up of any chemical moieties on surfaces [10].

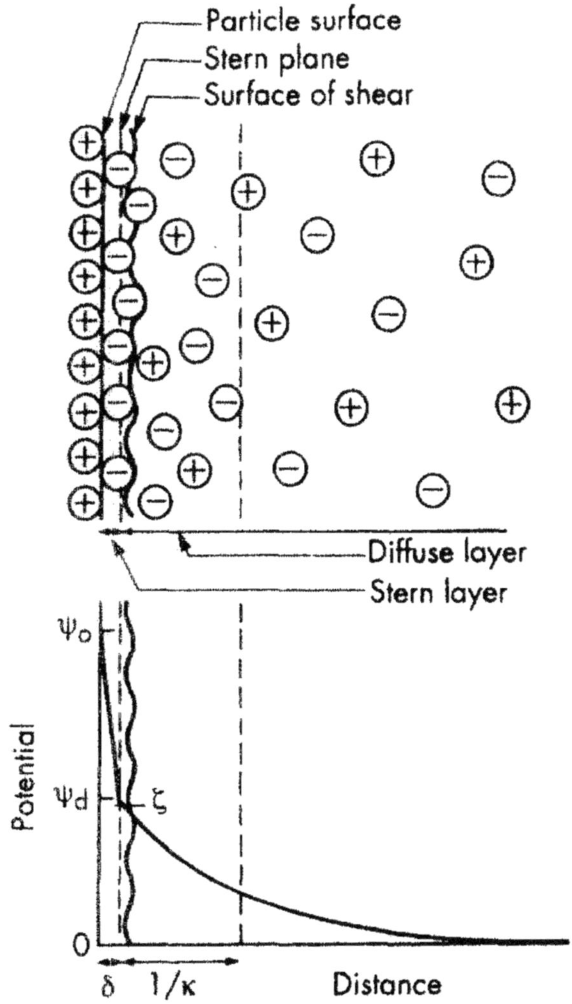

Fig. 4.2 Schematic representation of the structure of the electric double layer according to Stern's theory. Ψ_d = Stern potential, ζ = Zeta potential which corresponds to the shear plane close to Stern plane. Reprinted with permission from [11]

and ions move together with the surface and they are separated from the rest of liquid by a slipping plane. Electric potential at the slipping plane is named *zeta potential* (ζ).

This prevalent picture of the electrical double layer is often modified by specific adsorption effects. For instance, Ba^{2+} cations adsorb on hematite particles with excess charge and anionic surfactants adsorb on negative silicate and cellulose surfaces. pH plays a major role in determining surface charge, since most surfaces show specific binding of H^+ or OH^-, even inert surfaces like hydrocarbons and fluorocarbons. For instance, water flowing through polyethylene or polytetrafluoroethylene tubing acquires excess positive charge due to OH^- adsorption (Fig. 4.3) [13].

Changing the solution pH is an important tool to change the interfacial charge. Adding acid usually increases solid surface charge while adding base makes the surfaces more negative [14], as seen in Fig. 4.4. On the other hand, measuring

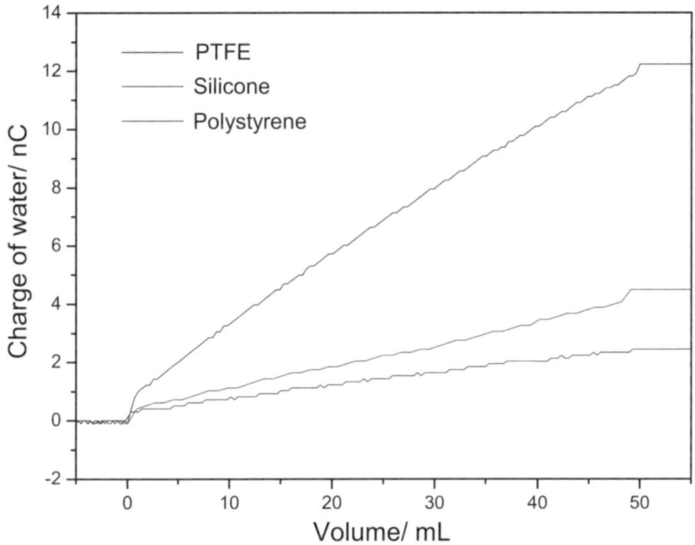

Fig. 4.3 Representative plots showing electrical charge acquired by 50 mL of deionized water after flowing through hydrophobic materials. Reprinted with permission from [13]

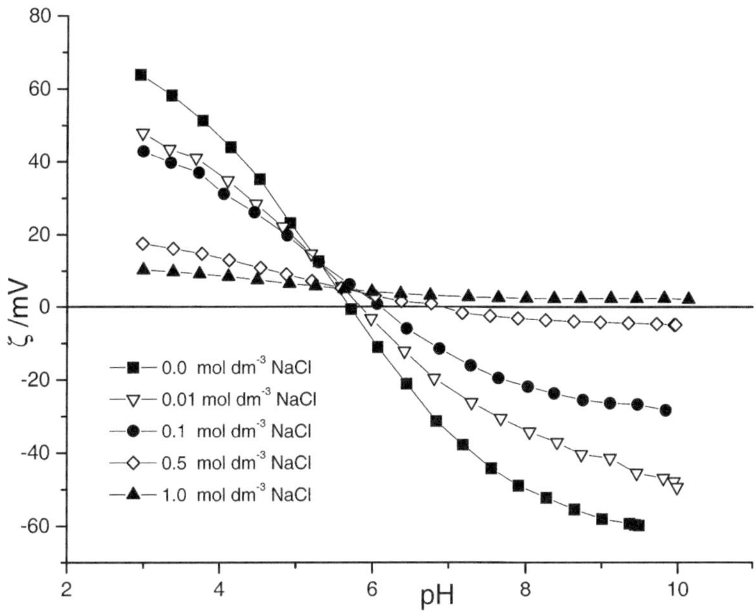

Fig. 4.4 Effect of pH for various NaCl concentrations on zeta potential of titanium dioxide (anatase). Reprinted with permission from [15]

the zeta potential of interfaces is an invaluable tool in the study of the chemical structure and properties of solid surfaces that is essential information, e.g. in mineral processing.

4.2.3 Experimental Methods

Potential measurements at the particle surface or at the limit of the Stern layer are not currently feasible. However, potential at the slipping plane is determined using electrokinetic data and it is called zeta potential. The main tools for assessing the electric properties of solid–liquid interfaces are thus potentiometric measurements using different types of electrodes and zeta potential measurements based on the various electrokinetic phenomena, especially electrophoresis, electrosmosis, and sedimentation potential. However, surface charge and electrokinetic charge are very different double layer characteristics and the significance of each should be clearly recognized, following a critical evaluation by Lyklema, who made outstanding contributions to this topic [16]. The availability of both types of data for well-defined systems makes accessible much relevant information on the electrical double layer, including counter-ion adsorption, stagnant layer conduction, and overcharging. Lyklema also discussed the interpretation in terms of specific adsorption and ion correlations.

Potentiometric techniques are apparently simple and easily implemented in a number of different systems. Unfortunately, this is often deceiving. Application of potentiometry to the determination of proton surface charge at mineral/water interfaces was recently reviewed [17], covering conventional experimental procedures and providing a critical discussion of problems with the techniques. Recommendations for obtaining reasonable and comparable results were made, discussing the most important experimental parameters. The authors proposed a reference titration procedure to allow comparisons of experimental data, providing a checklist for researchers and reviewers that could improve the usefulness of published data.

Other techniques for ion adsorption measurements are also used to determine surface charge. However, simple comparison of adsorption and potentiometry results may lead to apparent disagreement that may indeed reveal previously unsuspected complexity of the surface. For instance, in a study on charge in soil particles the authors observed that "...estimates of net surface charge by potentiometric titration and ion adsorption did not agree, especially as the soil pH was adjusted away from the point of zero charge. This lack of agreement is attributed to dissolution reactions of minerals and organic matter at high and low pH, which consume acid or base and overestimate surface charge [18]."

Electrophoresis is by far the richest source of zeta potential data but it is unsuitable to the study of macroscopic S/L interfaces, like glass or textiles. These can be comminuted into small particles for electrophoresis experiments but comminution is often accompanied by mechano-chemical reactions that produce new surfaces, different from those existing in the original solid. In this case, zeta potentials are calculated from electro-osmosis or streaming potential experimental results [19].

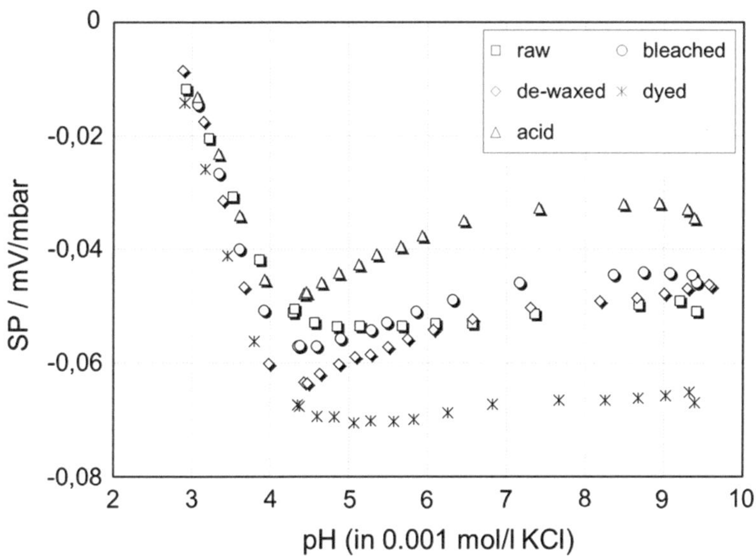

Fig. 4.5 Streaming potential of raw cotton fiber and the same but following various treatments. Reprinted with permission from [22]

Cellulose is a rather intractable polymer due to the scarcity of solvents and to the fact that it decomposes thermally prior to melting. For this reason, cellulose and cellulosic materials are interesting examples of the use of streaming potentials (Fig. 4.5) to characterize charged groups in a complex, fibrous, and hardly soluble solid, including film membranes [20–22].

4.3 Liquid–Liquid Interfaces

Potential difference is observed across liquid–liquid interfaces following different mechanisms. An important case is differential ion diffusion across the interface that produces junction potentials (see Chap. 2). Another type of frequently observed phenomena is selective ion adsorption, as in the so-called membrane electrodes and in negative charge formation at water–oil interfaces.

The latter is the important case of OH^- ion accumulation at the interfaces between water and non-polar media like oils [23] and polyolefins. It is also related to the complex behavior or ions at water–air interfaces that will be discussed in the forthcoming section.

4.4 Solid- and Liquid–Gas Interfaces

This is a broad topic covering many different situations, depending on the nature of the solid or liquid. The possibility of charge transfer to and from the atmosphere, mediated by water droplets and vapor, was recognized by Faraday in the "vapor electricity" discovered by Lord Armstrong and by Kelvin in his "water drop equalizer". This instrument was used to measure atmospheric electricity and his functioning was based on charge transfer between a water tank and the atmosphere, at a fast rate. According to Kelvin, ". . . any difference of potentials between the insulated conductor and the air at the place where the stream from the nozzle breaks into drops is done away with at the rate of five per cent, per half second, or even faster" [24].

However, this information was later neglected, e.g. in E. Schrödinger's doctoral thesis [25], and other work [26, 27] that assigned the effect of air humidity on electrostatic charge dissipation to the increased surface conductance of the electrified solids. The dominating role of water vapor as a charge carrier to and from S/L interfaces and the role of the atmosphere as a charge reservoir received additional support, recently [28].

4.4.1 Liquid–Gas Interfaces

Spontaneous establishment of electric potential difference at the water–air interface is historically very important since it was already used by Kelvin.

There are at least three other areas of investigation revealing important information on the electrical properties of liquid–gas interfaces: Langmuir-Blodgett monolayers, interfaces of aqueous electrolyte solutions, and the study of water carrying excess charge. The two first topics are discussed in this chapter while the latter is further discussed in Chaps. 6 and 7, in this book.

4.4.1.1 LB Monolayers

Langmuir-Blodgett monolayers [29, 30] have been widely used to study structure and properties of many systems, ranging from surfactants to viruses and quantum dots. They are also used to manufacture self-assembled films and multilayers. For these reasons, they are by themselves a huge topic where the importance of interfacial electric fields in phenomena like molecular interactions, reorientation, and induced conformational changes was clearly demonstrated [31] and this is often easier than is a 3D system. On the other hand, interfacial charge in monolayers is largely dependent on the complex monolayer-forming species rather than on more strictly interfacial phenomena. For this reason, this topic will not be treated in detail here and the interested reader is directed to the excellent books and reviews on this topic.

4.4.1.2 Interfaces of Water and Aqueous Electrolyte Solutions

Experiments [32–36], theoretical calculations [37, 38], modeling, and simulation [39, 40] done on these interfaces have produced an impressive amount of work [41], benefiting from new, sophisticated techniques [42] together with classical methods.

This section does not give a full account of all the work done in this area but it presents some examples of the experimental results addressing the main points of divergence among the various authors.

For instance, vibrational sum frequency spectroscopy has been a rich source of information on ion accumulation at L/G interfaces [43, 44], showing also their effect on the surface water structure. HCl, HBr, and HI cause a significant disruption in the hydrogen-bonding network at the air–liquid interface, similar to that which is observed for sodium halides. One of the observed effects is a decrease in the number of dangling OH bonds relative to the neat water surface. These authors also found evidence for the presence of hydronium and Zundel ions and for increased concentration of bromide and iodide ions at the interface, in agreement with the electron spectroscopy results.

One open question is especially relevant: is the surface of pure water acidic or basic [45, 46], this means, which ions are preferentially adsorbed at the water–air interface, H^+ or OH^-?

Garett [47] reviewed this topic emphasizing the accumulation of some anions at the air/water interface and not in the bulk, as usually happens to the cations. He also presented simulations explaining those positive surface adsorption excesses.

Jungwirth [33] used results from surface-selective spectroscopies and molecular simulations. Both approaches indicate that the heavier and thus softer halide ions can be present and even enhanced at the water surface. This author finds that hydronium but not hydroxide accumulate at the air/water and alkane/water interfaces. These findings were extended to water–protein interfaces and they supported a local model of interactions of ions with proteins aiming to the rationalization of ion-specific Hofmeister effects in the salting out of proteins.

An interesting analysis of the situation was made by Manciu and Ruckenstein [48], who applied a Poisson-Boltzmann model compatible with zeta potential experiments [49–52] to calculate the concentrations of ions at the interfaces. Their conclusions are as follows. For most pH values, the concentrations of both H^+ and OH^- are much larger in the interfacial region than in bulk. However, at very low pH values, the interfacial concentration of H^+ is lower than in bulk. At large pH values, the interfacial concentrations of both H^+ and OH^- are lower than in bulk for three independent sets of ion adsorption parameters, obtained from three different sets of experimental zeta potential results. These lower interfacial concentrations of H^+ and OH^-, at extreme pH values, arise from adsorption saturation due to the finite number of adsorption sites. The ratio between H^+ and OH^- concentrations is lower at the interface than in the bulk in most but not all cases. The interfacial ratio of H^+ and OH^-, divided by the same ratio in the bulk, depends on both the pH and salt concentration.

The thermodynamic model presented by Kallay and co-workers leads to some conclusions convergent with those in the previous paragraph, supporting experimental results on the dependency of the surface potential at the gas–water interface with the pH and showing that the interface is negative above pH 3.8 [53].

Further progress in this area is needed, not only to settle the current divergences but mainly because understanding these interfaces is needed to understand other important phenomena, like proton transfer at interfaces that is quite different from conventional proton transfer in bulk water [54]. The discrepancies between experimental results probably arise from the difficulty to avoid changing the excess charge in liquids and solids, inadvertently. This is probably due to handling or induction by neighboring solids, liquids, and the surrounding atmosphere that is often unduly neglected. The discrepancies among theoretical and computational results may arise from intrinsic limitations of the methods used or from computational difficulties to handle sufficiently large sets of molecules and ions.

4.4.2 Metal or Semiconductor/Liquid Interfaces

Metal or semiconductor–liquid interfaces are an essential topic of electrochemistry and electroanalytical chemistry, treated extensively in the excellent books authored by Bard [55] and Bockris [56]. A unique feature of electrodes is that the potential on the metal can be measured precisely with high time resolution. Moreover, it can also be changed by the researcher within a broad range, limited only by the onset of reduction and oxidation reactions at the two electrodes and the kinetics of the electrode reactions. The electrical double layer is a matter of great interest to electrochemists, who may count on a number of precise, time-resolved techniques whose power is not matched by the techniques available to colloid and surface chemists. For this reason, knowledge on the dynamic characteristics, structure and reactivity at electrode interfaces usually exceeds that available for many other kinds of interfaces [57].

A huge number of interesting and often unpredictable facts has been disclosed by electrochemical research, evidencing fascinating features of the electrode interfaces and of the power of electrochemical methods.

For instance, electrode interfaces can be used to generate unstable species, like radical ions in solution, by oxidizing or reducing stable precursors like polycyclic hydrocarbons and various nitrogenated compounds: aromatic amines, nitrocompounds, and nitrites. These free-radicals in solution engage in electron-transfer reactions that produce electrogenerated chemiluminescence (ECL), a powerful tool for ultrasensitive biomolecule detection and quantification using miniaturized biosensors capable of multiplexing detection with high sensitivity, low detection limit, and good selectivity and stability [58]. When ECL is produced within thermoresponsive redox microgels, its intensity is correlated with the collapse of microgel particles [59]. This is a good example of the flexibility in the access to complex, unstable species and fine control that are achieved by triggering chemical reactions at the metal–liquid interfaces.

4.5 Solid–Solid Interfaces

Potential differences develop across solid–solid interfaces, due to different mechanisms: (1) electron transfer, in metals and semiconductors; (2) transfer of tribo-ions formed mechanochemically; (3) selective partition of adsorbed ions, especially OH^-.

The first case is relevant in many areas, as in electrochemistry, corrosion, and microelectronics, and it is a well-developed topic that will not be further discussed here. The transfer of tribo-ions formed mechanochemically that was discovered recently is the topic of Chap. 9 in this book.

4.5.1 *Selective Partition of Adsorbed Ions*

Selective ion adsorption plays an important role that was first raised by the Whitesides group [60, 61], who described the fabrication and characterization of ionic electrets. These materials contain a long-lived electrostatic charge due to an imbalance between the number of cationic and anionic charges in the material. For instance, crosslinked polystyrene microspheres that contain covalently bound ions and mobile counterions transfer some of their mobile ions in air, in the absence of bulk liquid, to another contacting material. This selective transfer of mobile ions yields microspheres with a net electrostatic charge. A typical charge density is 1 elementary charge per 2000 nm^2, close to the theoretical limit imposed by the dielectric breakdown of air. It increases in an atmosphere of SF_6, compared to N_2. Other ionic electret materials are functionalized glass or silicon with covalently bound ions and mobile counterions. Charge patterns are built in these materials, using soft lithography [62].

These ideas have been extended to many other situations, because most solid surfaces contain some adsorbed water, even under low relative humidity, and water produces H^+ and OH^-. Moreover, the equilibrium concentrations of these two ions in pure water depend on the local electrostatic potential, under a non-zero electric potential. Following the presentation in Chap. 5, $[H^+]$ under a potential ϕ_1 is given by

$$-2.30RT\left(\log\left[H^+\right]-7\right)=F\left(\phi_1\right)$$

Thus, water in a region in space where $\phi_1 = 0.2\ V$ should have pH = 10.4. This is not macroscopically observed because the electric double layer formed at the aqueous interface shields the potential. Nevertheless, pure water contains excess concentration of H^+ or OH^- ions, depending on the adjacent electrostatic patterns so that it can supply ions for adsorption, depending not only on chemical affinity but also on the local potential.

4.6 Water Structures at Interfaces

The behavior of water at interfaces has been receiving great attention for many decades [63] with plenty of diverging results and conclusions drawn by distinguished authors. This is probably one of the topics where greatest disagreement is found, in current science. New proposals further contribute to the complexity of the situation and the level of conflict seems to increase continuously, adding to the many disputes that were already mentioned in this chapter. This is not surprising, considering that water itself is "The Most Anomalous Liquid" [64].

References

1. Iwamoto M (2012) Maxwell–Wagner effect. In: Bhushan B (ed) Encyclopedia of nanotechnology. Springer, Berlin, p 1276
2. Costa RA, Coltro L et al (1990) A staining procedure for the detection of oxidized sites in polyolefins. Angew Makromol Chem 180:85–94
3. Borukhov I, Andelman D et al (2000) Polyelectrolyte titration: theory and experiment. J Phys Chem B 104:11027–11034
4. Shannon RD (1976) Revised effective ionic radii and systematic studies of interatomic distances in halides and chalcogenides. Acta Crystallogr Sect A 32:751–767
5. Nightingale ER Jr (1959) Phenomenological theory of ion solvation. effective radii of hydrated ions. J Phys Chem 63(9):1381–1387
6. Noyes RM (1962) Thermodynamics of ion hydration as a measure of effective dielectric properties of water. J Am Chem Soc 84(4):513–522
7. Zhang R, Somasundaram P (2006) Advances in adsorption of surfactants and their mixtures at solid/solution interfaces. Adv Colloid Interf Sci 123–126:213–229
8. Manne S, Cleveland JP et al (1994) Direct visualization of surfactant hemimicelles by force microscopy of the electrical double layer. Langmuir 10:4409–4413
9. Ballesteros-Gómez A, Rubio S (2009) Hemimicelles of alkyl carboxylates chemisorbed onto magnetic nanoparticles: study and application to the extraction of carcinogenic polycyclic aromatic hydrocarbons in environmental water samples. Anal Chem 81:9012–9020
10. Somasundaran P (2002) Simple colloids in simple environments explored in the past, complex nanoids in dynamic systems to be conquered next: some enigmas, challenges, and strategies. J Colloid Interf Sci 256(3–15):3–15
11. Shaw DJ (1992) Introduction to colloid and surface chemistry. Charged interfaces, chapter 7. Butterworth Heinemann, London, p 174
12. Lyklema J (1995) Fundamentals of interface and colloid science: solid interfaces. Electric double layers, vol 2, chapter 3. Academic, New York, pp 3-1-3-232
13. Burgo TAL, Galembeck F et al (2016) Where is water in triboelectric series? J Electrostat 80:30–33
14. Quast K (2012) Effects of pretreatments on the zeta potential characteristics of a hematite ore. Int J Mining Eng Miner Process 1(2):47–55
15. Gustafsson J, Mikkola P, Jokinen M, Rosenholm JB (2000) The influence of pH and NaCl on the zeta potential and rheology of anatase dispersions. Colloid Surf A Physicochem Eng Asp 175:349–359
16. Lyklema J (2011) Surface charges and electrokinetic charges: distinctions and juxtapositionings. Colloid Surf A 376:2–8
17. Lützenkirchen J et al (2012) Potentiometric titrations as a tool for surface charge determination. Croat Chem Acta 85(4):391–417

18. Marcano-Martinez E, McBride MB (1989) Comparison of the titration and ion adsorption methods for surface charge measurement in oxisols. Soil Sci Soc Am J 53(4):1040–1045
19. Jacobasch HJ, Bauböck G et al (1985) Problems and results of zeta-potential measurements on fibers. Colloid Polym Sci 263:3–24
20. Werner C, Jacobasch HJ et al (1995) Surface characterization of hemodialysis membranes based on streaming potential measurements. J Biomater Sci Polym Ed 7(1):61–76
21. Hubbe MA (2006) Sensing the electrokinetic potential of cellulosic fiber surfaces. BioResources 1(1):116–149
22. Luxbacher T, Čurlin M et al (2014) Assessing the quality of raw cotton knitted fabrics by their streaming potential coefficients. Cellulose 21:3829–3839
23. Marinova K, Alargova R et al (1996) Charging of oil-water interfaces due to spontaneous adsorption of hydroxyl ions. Langmuir 12:2045–2051
24. Applin KL, Harrison RG (2013) Lord Kelvin's atmospheric electricity measurements. Hist Geo Space Sci 4(2):83–95. http://centaur.reading.ac.uk/33822/1/lord_Kelvins.pdf. Accessed 28 Jun 2016
25. Schrödinger E (1910) Über die Leitung der Elektrizität auf der Oberfläche von Iso-latoren an feuchter Luft. Ph.D. thesis, University of Wien, Wien
26. Blacker RS, Birley AW (1991) Electrostatic charge occurrence, significance and measurement. Polym Test 10(4):241–262
27. Field RF (2011) The formation of ionized water films on dielectrics under conditions of high humidity. J Appl Phys 17(5):318–325
28. Burgo TAL, Rezende CA et al (2011) Electric potential decay on polyethylene: role of atmospheric water on electric charge build-up and dissipation. J Electrostat 69:401–409
29. Knobler CM, Schwartz DK (1999) Langmuir and self-assembled monolayers. Curr Opin Colloid Interf Sci 4:46–51
30. Oliveira ON Jr (1992) Langmuir-blodgett films—properties and possible applications. Brazil J Phys 22(2):60–69
31. Nobre TM, Silva HS, Leone FA, Miranda PB, MED Z (2009) Molecular view of the interaction between t-carrageenan and a phospholipid film and its role in enzyme immobilization. J Phys Chem B 113:7491–7497
32. Healy TW, Fuerstenau DW (2007) The isoelectric point/point-of zero-charge of interfaces formed by aqueous solutions and nonpolar solids, liquids, and gases. J Colloid Interf Sci 309:183–188
33. Jungwirth P (2009) Spiers Memorial Lecture: ions at aqueous interfaces. Farad Discuss 141:9–30
34. Creux P, Lachaise J et al (2009) Strong specific hydroxide ion binding at the pristine oil/water and air/water interfaces. J Phys Chem B 113:14146–14150
35. Beattie JK, Gray-Weale A (2012) Oil/water interface charged by hydroxide ions and deprotonated fatty acids: a comment. Angew Chem Int Ed 51:12941–12942
36. Ghosal S, Hemminger JC et al (2005) Electron spectroscopy of aqueous solution interfaces reveals surface enhancement of halides. Science 307:563–566
37. Kudin KN, Car R (2008) Why are water-hydrophobic interfaces charged? J Am Chem Soc 130:3915–3391
38. Levin Y (2008) Polarizable ions at interfaces. Phys Rev Lett 102:147803
39. Tobias DJ, Stern AC et al (2013) Simulation and theory of ions at atmospherically relevant aqueous liquid-air interfaces. Annu Rev Phys Chem 64:339–359
40. Netz RR (2004) Water and ions at interfaces. Curr Opin Colloid Interf Sci 9:192–197
41. Jungwirth P, Winter B (2008) Ions at aqueous interfaces: from water surface to hydrated proteins. Annu Rev Phys Chem 59:343–366
42. Fayer MD (2012) Dynamics of water interacting with interfaces, molecules, and ions. Acc Chem Res 45:3–14
43. Levering LM, Roxana Sierra-Hernández M et al (2007) Observation of hydronium ions at the air-aqueous acid interface: vibrational spectroscopic studies of aqueous HCl, HBr, and HI. J Phys Chem C 111:8814–8826

44. Liu D, Ma G et al (2004) Vibrational spectroscopy of aqueous sodium halide solutions and air–liquid interfaces: observation of increased interfacial depth. J Phys Chem B 108:2252–2260
45. Beattie JB, Djerdjev AM et al (2014) pH and the surface tension of water. J Colloid Interf Sci 422:54–57
46. Beattie JB, Djerdjev AM et al (2009) The surface of neat water is basic. Farad Discuss 141:31–39
47. Garett BC (2004) Ions at the air/water interface. Science 303(5661):1146–1147
48. Manciu M, Ruckenstein E (2006) Ions at the air/water interface. J Colloid Interf Sci 304:541–544
49. Manciu M, Ruckenstein E (2003) Specific ion effects via ion hydration: I. Surface tension. Adv Colloid Interf Sci 105:63–101
50. Manciu M, Ruckenstein R (2004) The polarization model for hydration/double layer interactions: the role of the electrolyte ions. Adv Colloid Interf Sci 112:109–128
51. Manciu M, Ruckenstein E (2005) On the interactions of ions with the air/water interfaces. Langmuir 21:11312–11389
52. Manciu M, Ruckenstein E (2012) Ions near the air/water interface. I: Compatibility of zeta potential and surface tension experiments. Colloids Surf A 400:27–35
53. Kallay N, Preocanin T et al (2015) Thermodynamic model of charging the gas/water interface. J Phys Chem C 119:997–1007
54. Mishra H, Enami S et al (2012) Anions dramatically enhance proton transfer through aqueous interfaces. PNAS 109:10228–10232
55. Bard AJ, Faulkner LR (2000) Electrochemical methods: fundamentals and applications. Wiley, New York
56. Bockris JO, Reddy AKN (1998) Modern electrochemistry 1: Ionics, electrodics. Springer, Berlin
57. Bockris JO (1993) Teaching the double layer. J Chem Educ 60(4):265–268
58. Miao W (2008) Electrogenerated chemiluminescence and its biorelated applications. Chem Rev 108:2506–2553
59. Pinaud F, Russo L et al (2013) Enhanced electrogenerated chemiluminescence in thermoresponsive microgels. J Am Chem Soc 135:5517–5520
60. McCarty LS, Winkleman A et al (2007) Ionic electrets: electrostatic charging of surfaces by transferring mobile ions upon contact. J Am Chem Soc 129:4075–4088
61. McCarty LS, Whitesides GM (2008) Electrostatic charging due to separation of ions at interfaces: contact electrification of ionic electrets. Angew Chem Int Ed 47:2188–2207
62. Xia Y, Whitesides GM (1998) Soft lithography. Annu Rev Mater Sci 28:153–184
63. Björneholm O et al (2016) Water at interfaces. Chem Rev 116:7698–7726
64. Pettersson LGM, Henchman RH et al (2016) Water—the most anomalous liquid. Chem Rev 116(13):7459–7462

Chapter 5
Charge Patterns, Charge Separation

Contents

5.1 Charge Patterns: From Molecules to Bulk Matter

Charge patterns are observed in matter in our environment, from neutral atoms, molecules and up. In the former, positive charge is concentrated in the dense nuclei while negative charge spreads over the electrosphere.

Neutral molecules also display complex charge and potential patterns, according to direct experimental observation and to the results of theoretical calculations. As an example, Fig. 5.1 shows the electrostatic potential in the outer regions of the molecule, 3-amino-5-hydroxypyridine [1]. This is how the latter is 'seen' by an approaching reactant, and it is a guide to the reactivity and non-covalent interactions of this molecule.

There is thus an asymmetry in the association between charge and mass in atoms, where positive charge is associated with heavier matter than negative charge. This asymmetry persists in some macroscopic systems like metals while in other cases the heavier part is negative, as in most silica and silicate particles [2] that make a

F. Galembeck, T.A.L. Burgo, *Chemical Electrostatics*,
DOI 10.1007/978-3-319-52374-3_5

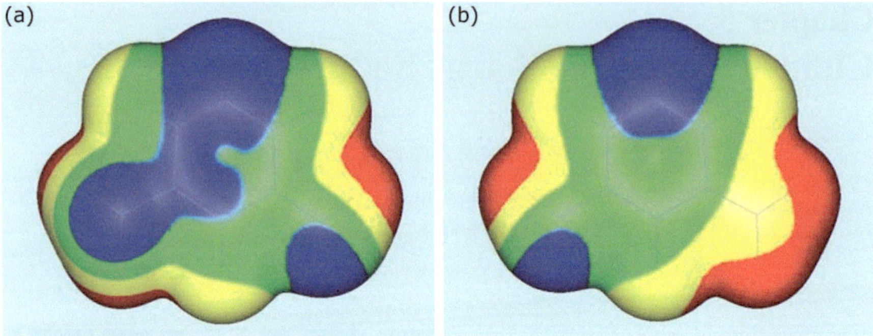

Fig. 5.1 Computed electrostatic potential on the molecular surface of 3-amino-5-hydroxypyridine. *Blue* is highly negative, *red* is highly positive. (**a**) Azine nitrogen is at the top, hydroxyl group at right, and amino group at left showing its lone pair (*blue*). (**b**) Azine nitrogen is at the top, amino group at right showing its hydrogens (*red*) and hydroxyl group, at left. Reprinted with permission from [1]

large part of the Earth's crust. These particles are formed by a stiff negative matrix formed by Si, O and eventually some metal ions, together with H^+ and dissociable metal counter-ions. Acknowledging these differences is important, for understanding the various mechanisms for charge separation and the coupling between mass and charge transfer.

Anisotropy is a frequent cause for charge separation that occurs in every interface, leading to charge separation by a number of different mechanisms. This will be discussed separately in Chap. 8 in this book.

The appearance of techniques for mapping electric charge and potential from macro- to nano-scale revealed pronounced contrast in most samples examined [3, 4], showing that local charge patterning is the rule in dielectrics (Figs. 5.2 and 5.3). Moreover, positive and negative charges usually have a non-zero sum.

5.2 Charge Separation Within Solids, Liquids and Gases

Simple descriptions of electrostatic charging in solids assign it to the transfer of electrons between contacting media due to the differences in work functions. However, the reality is much richer than this.

This section describes basic phenomena that contribute to the multiplication of charged species that will finally separate producing domains with excess charge and thus charge and potential patterns.

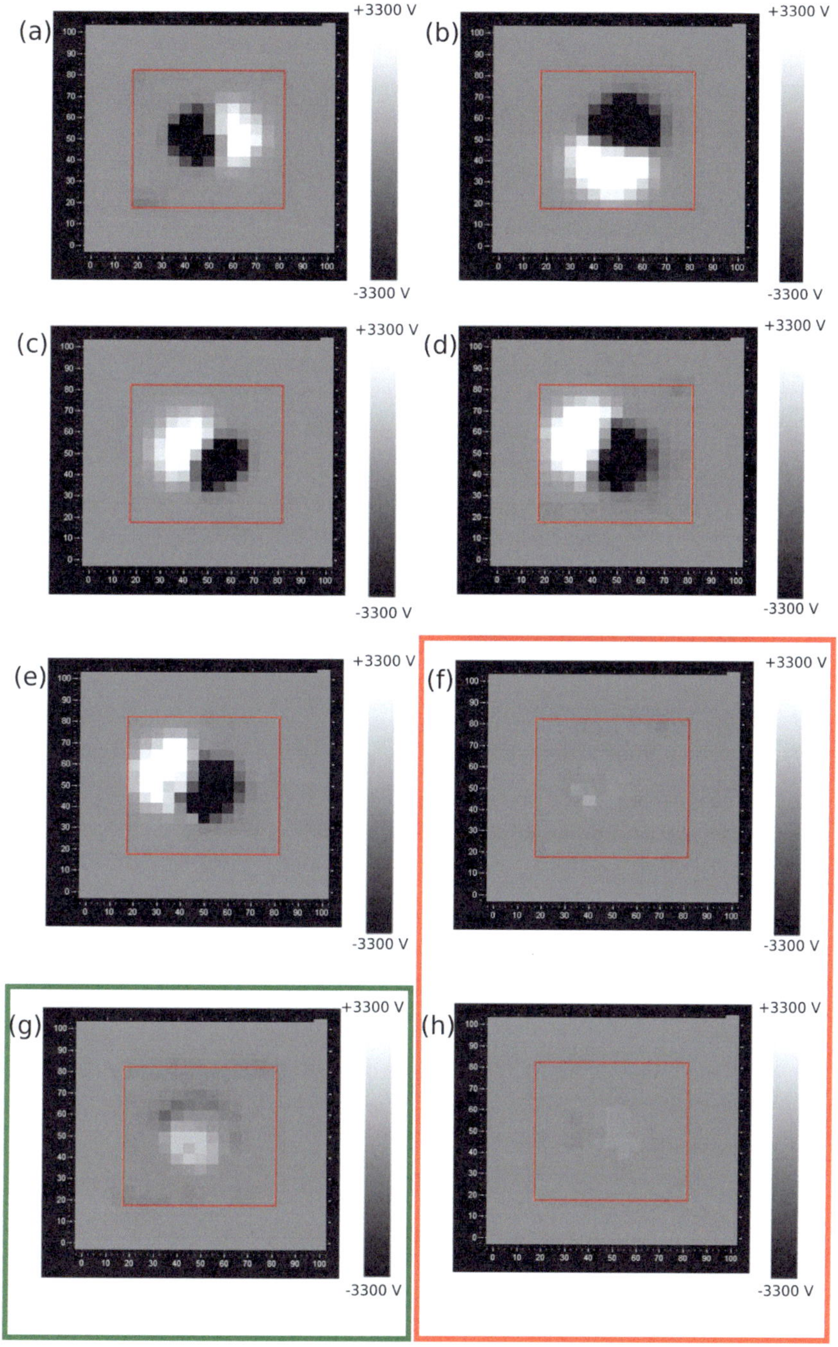

Fig. 5.2 Electric potential maps of many polytetrafluoroethylene (PTFE) and polyethylene (PE) samples. (**a**–**e**) PTFE rubbed with PE foam shows a reproducible macroscopic planar dipole, (**f**, **h**) high-density polyethylene (HDPE) film and (**g**) PE foam rubbed with PTFE show positive and negative domains only with much lower potentials than PTFE. Tribocharging times varied between 1 and 3 s. Reprinted with permission from [4]

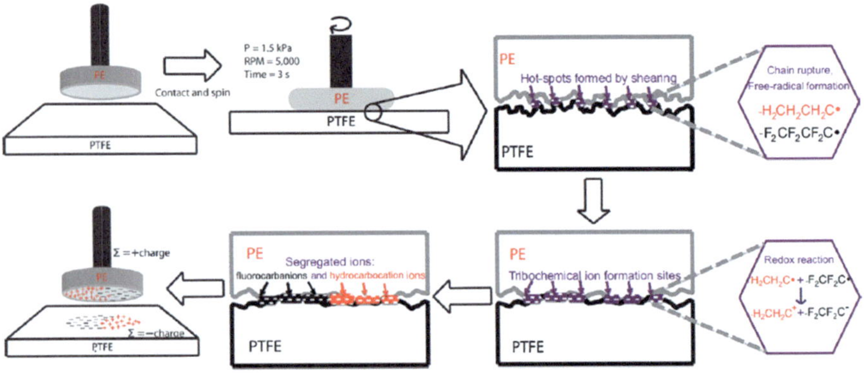

Fig. 5.3 Mechanism for contact triboelectrification of insulating polymers. Shearing the polymer interface heats both surfaces unevenly forming hot spots, due to forced contact on surface hills. Plasticization and melting take place, added to chain breakdown and fragmentation. Homolytic scission produces free radicals with markedly different electronegativities that are converted into fluorocarbanions and hydrocarbocations by electron transfer. Ions are segregated due to the chain size, following Flory-Huggins theory and superseding weak electrostatic interactions between highly spaced charges. Reprinted with permission from [4]

5.2.1 Ionization and Ion Separation

Neutral molecules like H_2SO_4 and acetic acid, CH_3COOH, undergo ionization when dissolved in water and other solvents and this is sometimes assigned to the large dielectric constant of water. However, these two acids also undergo self-ionization, in the absence of solvent, like many other substances including water itself [5].

$$2\mathrm{H_2O} \rightleftharpoons \mathrm{H_3O^+} + OH^-$$

$$2NH_3 \rightleftharpoons NH_4^+ + NH_2^-$$

$$2\mathrm{H}_2SO_4 \rightleftharpoons \mathrm{H}_3SO_4^+ + HSO_4^-$$

$$3HF \rightleftharpoons \mathrm{H_2F^+} + HF_2^-$$

$$2PF_5 \rightleftharpoons PF_6^- + PF_4^+$$

$$2\mathrm{N_2O_4} \rightleftharpoons NO^+ + NO_3^-$$

$$2BrF_3 \rightleftharpoons BrF_2^+ + BrF_4^-$$

Ions are often formed in reactions between neutral molecules, as in the preparation of halonium salts [6]. The addition of methyl bromide or methyl chloride in sulfur dioxide at −78 °C to a complex of antimony pentafluoride and tetrafluoromethane in sulfur dioxide yields crystals of $[CH_3–X]^+$ $[CH_3SbF]^-$, following the evaporation of SO_2. This salt is stable at room temperature but it is sensitive to water.

Superacids [7] like carborane, $H(CHB_{11}Cl_{11})$, are important cases of substances showing a great stability in the ionized state. Carborane is an acid one million times stronger than sulfuric acid, due to the extreme stability of the $CHB_{11}Cl_{11}^-$ anion. Carborane and other superacids are extreme cases of the asymmetry between charge and mass distribution, given the large difference in the sizes of H^+ and the anions that leads to large differences, for instance, in transport properties of the anion and cation.

The tendency to ionization in neutral molecules and molecular bonds derives from the differences in electronegativity of the participating elements, following the conceptual valence-bond framework consolidated by Pauling [8].

5.2.1.1 Solvation

Solvation plays an important role in the ionization status of solutes. A dramatic example used in Physical Chemistry college teaching is the case of NaCl, where ion solvation by water plays an essential role in reaching a negative Gibbs energy for the formation of solutions from the elements (and water) [9].

This topic was recently reviewed by Cox [10], considering electrolytes and non-electrolytes in a wide range of solvents and showing that the dissociation constant of an acid changes with solvent in a direct and quantitative relationship with the solvation energies of all the species involved. For instance, the transfer of simple anions from water to non-aqueous solvents is generally disfavored because of the loss of the contribution of hydrogen-bond interactions with water. On the other hand, cation solvation is highly favorable in solvents such as dimethylsulfoxide, which are strong Lewis bases, but low-polarity and low-basicity solvents such as acetonitrile and tetrahydrofuran do not make a contribution comparable to water. Non-electrolytes, such as carboxylic acids and amines, are more stable in non-aqueous media and this is clearly observed for fatty acids and amines, but the solvent effects are not as pronounced as in the case of ions. In mixed solvents, preferential solvation by the most favorable solvent component determines the dependence of solvation and dissociation constants upon solvent composition.

Great progress in this area was obtained thanks to the introduction of ion-cyclotron resonance that allowed the acquisition of acidity data in the gas phase [11]. A surprising result was the finding that toluene is more acidic than water in the gas phase, but ca. 20 orders of magnitude less acidic in solution. The reactions of toluene with water follow opposite paths in the gas phase and in water: in the former, toluene reacts with OH^- ion yielding $C_6H_5CH_2^-$ and H_2O, while in water the benzyl anion reacts with water, forming toluene and OH^-. Also, malononitrile in the gas phase is an acid stronger than acetic acid.

Clearly, ionization is not just an intrinsic feature of a molecule but it depends largely on its environment. This section showed how important are the effects of solvent and gas phase. The effect of interfaces will be addressed in a separate chapter.

5.2.1.2 Dielectric Rupture, Breakdown Voltage

Neutral atoms and molecules undergo charge separation under electric fields or when they are hit by background electrons accelerated under an electric field. This is widely used, e.g. in fluorescent and other electrical discharge lamps and mass spectroscopy. On the other hand, it also originates unwanted electrostatic discharge and sparkles that may trigger fire.

Ionization is the triggering step of a dielectric breakdown event but many other factors are also important in different situations. In gases, breakdown voltage that is the voltage necessary to start a discharge or electric arc between two electrodes depends on the pressure and gap length, represented by Paschen's law [12].

A special case of electrical breakdown is corona discharge produced around a sharp electrified tip, where the strength of the electric field is maximum. It often produces a glow and a sizzling sound, not only in intentionally electrified equipment but also in tree tops, ship masts and sharp ends in buildings. Corona discharge is used in ozone production and in polymer surface treatment when the treated materials acquire excess charge. This is used to make electrets [13] that are discussed in Chap. 10.

5.2.1.3 Ionizing Radiation

Ionizing radiation carries sufficient energy to extract electrons from atoms, molecules, particles and macroscopic matter. The required minimum energy also depends on the chemical substance involved, since ionization potentials for the elements range from 3.9 eV for cesium metal to 24.6 eV for helium.

The relevant magnitude in solids is the work function that ranges between 2.1 for Cs and 5.9 for Se, Os and Pt, considering only metals and semi-metals. The work function depends not only on the involved element but also on the solid surface properties including its contaminants. A striking verification of these ideas is plasmon-induced electron transfer from gold nanodots into TiO_2 [14] allowing the separation of charges under visible light radiation, and thus increasing the photocurrent in TiO_2 films [15]. This follows previous developments on the photosensitization of wide-band gap nanocrystalline semiconductors by adsorbed dyes [16], bringing the radiation energy level required for charge separation and electron withdrawal down to the center of solar spectrum at Earth surface.

On the other extreme of the energy scale, cosmic rays are high-energy particles impinging on Earth and provoking air showers [17], this means, extensive cascades of ionized particles and electromagnetic radiation. It has been suggested that cosmic-ray-induced air showers in combination with runaway breakdown may initiate lightning [18].

5.2.1.4 Mechanochemistry, Tribochemistry

Mechanical action on insulators provokes the appearance of a triboplasma [19, 20] containing a large number of non-equilibrium, high-energy species including free radicals and ions. The primary species decay to more stable entities producing many species that are not obtained by thermal or electrochemical reactions but have some similarity with the products of radiochemical reactions. The involved phenomena, mechanisms, and their consequences are treated in Chap. 8, in this book.

5.2.2 Charge Segregation

The previous section showed that ionized matter is widespread, under equilibrium and non-equilibrium conditions. A related question concerns the separation of charged components of matter in domains carrying excess positive or negative charge and the thermodynamic or kinetic stability of these domains. The following sections describe two main groups of mechanisms for charge segregation: differential mass transfer and self-assembly.

5.2.2.1 Differential Mass Transfer

There is no correlation between the mass of ions, charged macromolecules and particles and their charge, large or small, positive or negative. There are also not any broad correlations between charge sign and amount and chemical composition, although some statements can be made, like "elements at the left of the periodic table form cations while halogens and chalcogens form anions".

This creates many possibilities for charge segregation triggered by mass transfer, under equilibrium or transient conditions. The two following sections describe the cases of sedimentation and streaming potential, liquid junction potential and membrane potential.

5.2.2.2 Sedimentation Potential and Streaming Potential

Sedimentation and streaming potentials are two electrokinetic phenomena depending on the formation of electric double layers at interfaces. This kind of charge separation takes place when water and other liquids contact charged particles, macromolecules or macroscopic solid or liquid bodies [21]. Charging at interfaces is treated in this book, in Chap. 6.

Sedimentation potential is also known as the Dorn effect, since it was first reported in 1879 by Dorn, who observed the appearance of a vertical electric field while glass beads were settling in water. Streaming potential is the electric potential developed when an electrolyte flows across a porous medium with charged walls and it was also first observed long ago by Quincke, in 1859.

Sedimentation potential was used by Tolman [22] to determine ion transport coefficients in electrolyte solutions. It was also used to understand the effect of low ionic strength on the low MW determination results for proteins using sedimentation-diffusion equilibria. In this case, the unshielded repulsion between charged protein molecules, for instance, prevents them from reaching the concentration gradients predicted by the sedimentation equilibrium equation.

Streaming potential is also important in research, e.g. in zeta potential determinations. In geology, it contributes to the *spontaneous potential* measured between two electrodes down boreholes for the evaluation of formations, for oil and gas production, mineral exploration, dam seepage and other groundwater investigation. The large amount of data logging in oil exploration probably makes the Earth crust the most intensively mapped environment, concerning electric potentials.

The existence of sedimentation potentials, by itself, shows that electroneutrality is not expected along any solution contained within a tube undergoing centrifugation. Simple extension of Tolman's equations to larger solutes or particles dispersed in a liquid shows that electroneutrality should not also be expected along the depth of any water body where charged particles (clays, sand, humic matter) are dispersed, settling under gravity. Recent contributions from Levine and Ohshima [23] developed a theory for concentrated suspensions of spherical particles.

Considering that sedimentation of charged particles and fluid motion through porous media are widespread phenomena in important natural environments like the atmosphere, oceans and the solid crust, we should expect contributions of sedimentation and streaming potentials to all the surrounding electric patterns on Earth's surface.

5.2.2.3 Liquid Junction Potential and Membrane Potential

Liquid junction potential (LJP) is a concept familiar to electrochemists, electrophysiologists [24] and to any person that uses a combined glass-reference electrode for pH measurement. It appears when a liquid–liquid junction is made, contacting two liquids with different concentration and chemical composition of electrolyte solutes [25], or even two different liquids [26].

It is easily understood considering that the diffusion coefficients of the cation and anion in an electrolytic solution are usually different, as well as their electrophoretic mobilities and transport numbers as shown in Fig. 2.3. The magnitude of junction potentials cannot be overlooked and it has been demonstrated that LJPs formed in microchannels can induce appreciable electrophoretic transport of charged species without the use of electrodes or an external power supply [27].

Membrane potential is observed whenever a membrane separates two electrolytic solutions. This is important in various natural and technological processes and it has been widely used in chemical analysis, in the glass electrodes [28] for pH and other ions measurement and various membrane electrodes that were developed for a large number of analytes [29]. Moreover, it is a basic concept in cell biology [30]: ions may flow across the membrane by active and passive mechanisms and the changes in the membrane potential trigger important physiological phenomena.

Liquid junction and membrane potential are components of the "spontaneous potential" that appears during drilling operations in the oil industry. It was one of the first logging measurements ever made and it was discovered by accident, since it caused perturbations on the electric logging systems. It records the naturally occurring voltage produced by the interaction of connate water, drilling fluid, and shale. Its usefulness was soon realized, and it remains as a useful logging measurement, after many years [31]. In this case, the "membrane" is a layer of shale or sand.

5.2.2.4 Self-Assembly

Self-assembly creates several mechanisms for charge segregation and this section will examine the interesting case of micelle formation.

Other important cases of the interplay between electrostatic and hydrophobic interactions are biological systems like proteins, nucleic acids, viruses and cell membranes, where complex charge patterns are part of elaborate 3D structures, contributing to their stability. This topic is treated in Chap. 10.

5.2.2.5 Supramolecular Structures, Micelle Formation

Amphiphilic substances dissolved in water form different types of supramolecular aggregates: spherical, cylinder, lamellar and other micelle types. This is a remarkable case of formation of clusters of positive or negative ions driven by hydrophobic interactions. Micelle charge is compensated by the overall charge of counter-ions distributed within the diffuse part of the double-layer [32].

5.3 Pattern Propagation

Electrostatic forces are long-range, as compared to all other types of intermolecular interactions. Moreover, they follow simple additive rules, represented by the superposition principle. Consequently, an electrified object creates non-zero electric potential at great distances as testified by hair-rising on top of tall buildings and mountains, when storms approach.

Electric potential modifies the chemical potential of ions, changing their reactivity and tendency to migrate. The chemical potential is an important thermodynamic quantity that carries nearly all the information required for making predictions on the transformations underwent by a chemical substance within any system, under given pressure, temperature and chemical composition [33].

The sum of the chemical potential and the electrical potential energy for any component of a system is the electrochemical potential:

$$\Delta\mu_i^{el} = \Delta\mu_i + z_i F \Delta\phi$$

This equation shows that applying a non-zero potential on any ion will increase its electrochemical potential, if both have the same sign. This will bring the ion to a non-equilibrium state and it will migrate away toward an area with lower potential, or it will react thus decreasing μ_i or both, to recover equilibrium.

These effects are rather pronounced in electrolyte solutions but they may also have a strong effect on auto-ionizable substances like pure water, including vapor and ice. Writing,

$$\Delta\mu_i^{el} = \Delta\mu_i + z_i F \Delta\phi$$

for two different states 1 and 2, $\Delta\mu_i{}^{el} = 0$ requires,

$$-RT\ ln\ \frac{a_{i,1}}{a_{i,2}} = z_i F\left(\phi_1 - \phi_2\right)$$

This means that a difference of potential equal to 58 mV leads to a change in the activity of ion *i*, by one order of magnitude, and vice versa. This is the basis for *membrane and action potentials* in biophysics [34], for potentiometric measurements including pH, for enhanced reactivity in micelle solutions [35] and it plays an important role in electrochemical kinetics.

Most often, this equation is applied to stable ions dissolved in water. However, water itself undergoes ionization, although to a rather limited extent expressed in the $pK_w = 14$. This means, pure water at room temperature contains equal concentrations of H^+ and OH^- ions, 10^{-7} mol/L. Water under positive potential is thus enriched in OH^- ions and depleted from H^+ ions, transforming itself in a domain with excess negative charge. This has been proven by producing water with excess positive or negative charge and by imparting net charge to hydrophilic solids, a topic presented in detail in Chaps. 6 and 8.

Thus, water is the essential agent of a powerful mechanism for the propagation of electrostatic patterns, in space and time, since it is found is most natural and anthropic Earth environments.

5.4 Stability and Decay Rates of Charge Patterns

5.4.1 Systems Under Equilibrium

Many systems under thermodynamic equilibrium display charge patterns at various size ranges, as for instance contacting metals, micellar solutions, water surfaces and Langmuir-Blodgett films. These charge patterns are persistent, provided their ambient does not change.

However, potential differences within these patterns seldom exceed few tenths of a volt. This is sufficient for producing large biological effects but still orders of magnitude lower than the potentials in patterns that create the most spectacular displays of electrostatic phenomena.

5.4.2 *Non-Equilibrium Systems*

Dielectric rupture, ionizing radiation, corona discharge and mechanochemical reactions are powerful agents of charge separation producing large electrostatic potential in many systems. Potential half-lives vary widely depending on various factors, especially the nature of charge together with electrical conductivity and relaxation mechanisms in the solid.

Solids carrying steady excess charge form one class of electrets that will be further discussed in Chap. 7.

The main acknowledged mechanism for potential dissipation in electrified solids is electric current flow to the ground, bringing the potential on the solid down to the reference potential. However, we recall that the "ground" itself is not really at zero potential (see Chap. 1). Recent work showed that moist atmosphere is also effective in dissipating charge (see Chap. 6).

References

1. Murray JS, Politzer PT (2011) The electrostatic potential: an overview. Wiley Interdiscip Rev Comput Mol Sci 1(2):153–163
2. Kobayashi M, Juillerat F, Galletto P, Bowen P, Borkovec M (2005) Aggregation and charging of colloidal silica particles: effect of particle size. Langmuir 21:5761–5769
3. Galembeck A, Costa CAR et al (2001) Scanning electric potential microscopy imaging of polymers: electrical charge distribution in dielectrics. Polymer 42:4845–4851
4. Burgo TAL, Ducati TRD et al (2012) Triboelectricity: macroscopic charge patterns formed by self-arraying ions on polymer surfaces. Langmuir 28:7407–7416
5. Housecroft C, Sharpe AG (2007) Inorganic chemistry. Prentice Hall, London
6. Olah GA, DeMember JR (1970) Friedel-Crafts chemistry. V. Isolation, carbon-13 nuclear magnetic resonance, and laser Raman spectroscopic study of dimethylhalonium fluoroantimonates. J Am Chem Soc 92(3):718–720
7. Olah GA, Prakash GK et al (2009) Superacid chemistry. Wiley-Intercience, Hoboken, NJ, p 41
8. Pauling L (1960) The nature of the chemical bond. Cornell University Press, Ithaca, NY, pp 98–100
9. Atkins P, Paula J (2014) Atkins' physical chemistry. Oxford University Press, Oxford
10. Cox BG (2013) Acids and bases: solvent effects on acid-base strength. Oxford University Press, Oxford
11. Reichardt C, Welton T (2011) Solvents and solvent effects. In: Organic chemistry. Wiley, New York
12. Wadhwa CL (2007) High voltage engineering. New Age Science, pp 10–12.
13. Kestelman VN, Pinchuk LS et al (2000) Electrets in engineering: fundamentals and applications. Springer, New York

14. Furube A, Du LH et al (2007) Ultrafast plasmon-induced electron transfer from gold nanodots into TiO_2 nanoparticles. J Am Chem Soc 129:14852–14853
15. Su YH, Ke Y et al (2012) Surface plasmon resonance of layer-by-layer gold nanoparticles induced photoelectric current in environmentally-friendly plasmon-sensitized solar cell. Light Sci Appl 1(e14):1–5
16. Wang Q, Ito S et al (2006) Characteristics of high efficiency dye-sensitized solar cells. J Phys Chem B 110:25210–22522
17. Kampert KH, Watson AA (2012) Extensive air showers and ultra high-energy cosmic rays: a historical review. Eur Phys J H 37:359–412
18. Gurevich AV, Antonova VP et al (2013) Cosmic rays and thunderstorms at the Tien-Shan mountain station. J Phys Conf Ser 409:012234
19. Heinicke G (1984) Tribochemistry. Carl Hanser Verlag, München–Wien
20. Baláž P (2008) Mechanochemistry in nanoscience and minerals engineering. Springer, Berlin
21. Lyklema J (1995) Fundamentals of interface and colloid science: solid–liquid interfaces, vol. 2. Academic, New York, p 3.208
22. Tolman RC (1911) The electromotive force produced in solutions by centrifugal action. J Am Chem Soc 33:121–147
23. Ohshima H (1998) Sedimentation potential in a concentrated suspension of spherical colloidal particles. J Colloid Interf Sci 208:295–301
24. Barry PH, Lynch JW (1991) Liquid junction potentials and small cell effects in patch-clamp analysis. J Membr Biol 121:101–117
25. Dryfe RAWI (2007) In: Zosky CG (ed) Handbook of electrochemistry. Elsevier Science, Amsterdam, pp 849–877
26. Bunakova LV, Khanova LA et al (2004) Water-solvent liquid junction potential for some low-dielectric solvents. J New Mater Electrochem Syst 7:241–245
27. Munson MS, Cabrera CR et al (2002) Passive electrophoresis in microchannels using liquid junction potentials. Electrophoresis 23:2642–2652
28. Graham DJ, Jaselskis B et al (2013) Development of the glass electrode and the pH response. J Chem Educ 90(3):345–351
29. Rover L, Garcia CAB et al (1998) Acetylsalicylic acid determination in pharmaceutical samples by FIA-potentiometry using a salicylate-sensitive tubular electrode with an ethylene-vinyl acetate membrane. Anal Chim Acta 366:103–109
30. Wright SH (2004) Generation of resting membrane potential. Adv Phys Educ 28(4):139–142
31. Glover PWJ (2015) Petrophysics MSc Course. Notes: the spontaneous potential log. Chapter 18. Department of Geology and Petroleum Geology, University of Aberdeen, UK, p 218. https://groups.google.com/forum/#!msg/msc13iitkgp/NgvmuYtZm6A/ESXFV8LfSAMJ. Accessed 20 Sept 2016
32. Fendler H, Fendler EJ (1975) Catalysis in micellar and macromolecular systems. Chapter 2. Academic, New York, p 19
33. Job G, Herrmann F (2006) Chemical potential—a quantity in search of recognition. Eur J Phys 27:353–371
34. Barnett MW, Larkman PM (2007) The action potential. Pract Neurol 7(3):192–197
35. Quina FH (2016) Modeling chemical reactivity in ionic detergent micelles: a review of fundamentals. J Brazil Chem Soc 27(2):267–277

Chapter 6
Hygroelectricity: The Atmosphere as a Charge Reservoir

Contents

6.1 Water Vapor Sorption in Solids

Water sorption and desorption are very important in nature, since the survival of any species largely depends on water availability. Wherever water is scarce, the atmospheric vapor becomes an important source of water. The success of a species in these places depends then on developing strategies for using condensed water or for adsorbing it from the air.

"Non-rainfall" water on soil is important in arid zones, where dew can be the sole source of water for plants. Dew formation and direct water vapor adsorption are

F. Galembeck, T.A.L. Burgo, *Chemical Electrostatics*,
DOI 10.1007/978-3-319-52374-3_6

effective mechanisms for supplying water to the soil. Depending on surface temperature, water vapor adsorption is the only mechanism for water uptake by the soil [1].

Water vapor adsorption on solids is also a matter of great practical interest for many important products like pharmaceuticals, food, clothing and electronic ware whose production, durability and performance may be strongly affected by adsorbed water.

6.1.1 Water Vapor Adsorption Isotherms

Water vapor adsorption is known among practitioners as a broad, complex topic. Its difficulties are indeed expected, considering the relative contributions of various types of intermolecular interactions leading to the tendency of water molecules to cluster [2]. Another important factor is the pronounced amphoteric character of water that strongly contributes to its interactions with many solids.

An important feature is the size of water molecule that is smaller than the widely used N_2. For this reason, it was used instead of nitrogen to determine the pore structure in solids many years ago, using BET isotherms and especially *t*-curves that yield the statistical thickness of water films. A main advantage is that many important adsorbents have a significant part of their pore systems inaccessible to nitrogen but accessible to water vapor [3].

Other isotherms were used for different systems. For instance the Dubinin-Radushkevitch equation [4] has been widely used to quantitatively describe the adsorption of gases and vapors by microporous sorbents, as for instance carbon. This equation is based on the assumption that pore-filling is the relevant mechanism, rather than multilayer surface coverage. In one recent work, the modified Dubinin–Astakhov equation was used in the cases of activated carbon and alumina while the Langmuir equation was used in the case of carbon impregnated with $CaCl_2$, yielding good agreement with the experimental data [5].

Beyond isotherms, the structural properties of adsorbed water are by themselves a challenging topic. This is illustrated by a recent publication [6] on water vapor adsorption on single-wall nanotubes: water molecules form a thin (1–2 monolayers) adsorption layer created by lateral hydrogen bonds among molecules bound to the nanotube by van der Waals forces. The authors did not find evidence for charge transfer and the nanotubes coated with water are macroscopically hydrophobic. Following this paper and a large part of the literature on water vapor adsorption, electrostatic interactions do not play a role in water adsorption or desorption at the solid–gas interfaces. However, this role has now been demonstrated in many cases that are described in the following sections of this chapter.

6.2 Charging Cellulose Under High Humidity

Conventional wisdom says that electrostatic charge tends to disappear under high atmospheric humidity. For this reason, the results reported by Soares et al. [7] were rather surprising and they are among the first recorded evidence showing charge build-up under *high* humidity.

This work used a simple but well defined set-up that produced highly reproducible experimental results. Essentially, the positive electrostatic potential of an electrified acrylic sheet was measured under controlled relative humidity (RH) values, using a Kelvin electrostatic voltmeter. A sheet of filter paper that is essentially a cellulose film was then quickly introduced between the electrified solid and the electrode, while the Kelvin instrument readings were recorded, decaying toward zero. Then, the acrylic sheet was quickly withdrawn and the voltmeter readings became negative, showing that the paper had accumulated excess negative charge under the positive potential of the acrylic. The negative potential of the paper which was dissipated decayed again back to zero, following a curve symmetrical to that observed in the first step of the experiment, when the acrylic sheet was introduced. This is shown in detail in Fig. 6.1.

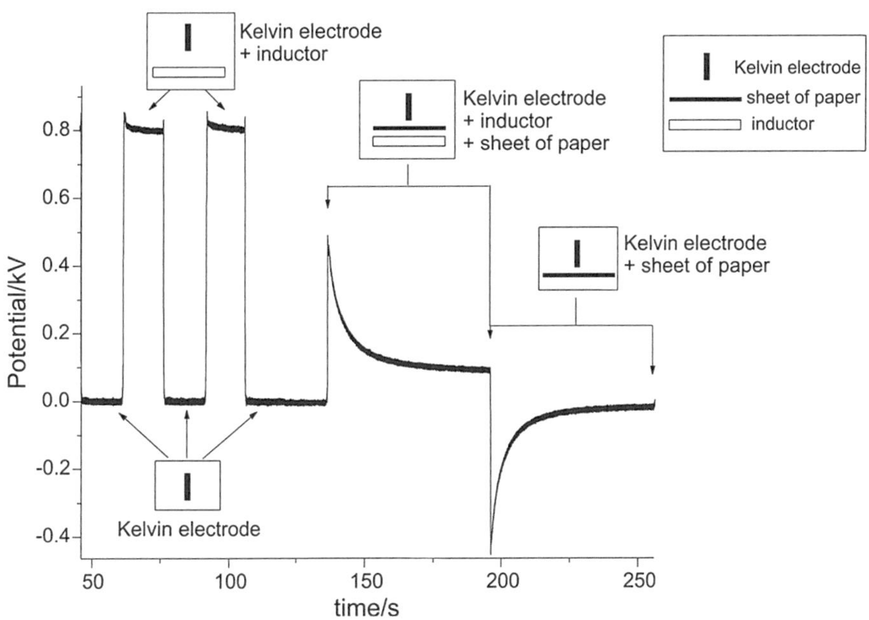

Fig. 6.1 Potential vs. time (at 10% RH and with one sheet of paper). The inductor was periodically introduced beneath the electrode during the experiment. The quasi-square waves to the left were recorded when the inductor was introduced and removed while the paper sample was withdrawn. The spikes in the central part show potential readings when the paper sample is positioned beneath the electrode. Positive potentials were observed when the inductor was introduced and near-zero or negative potentials were obtained while the inductor was withdrawn. Reprinted with permission from [7]

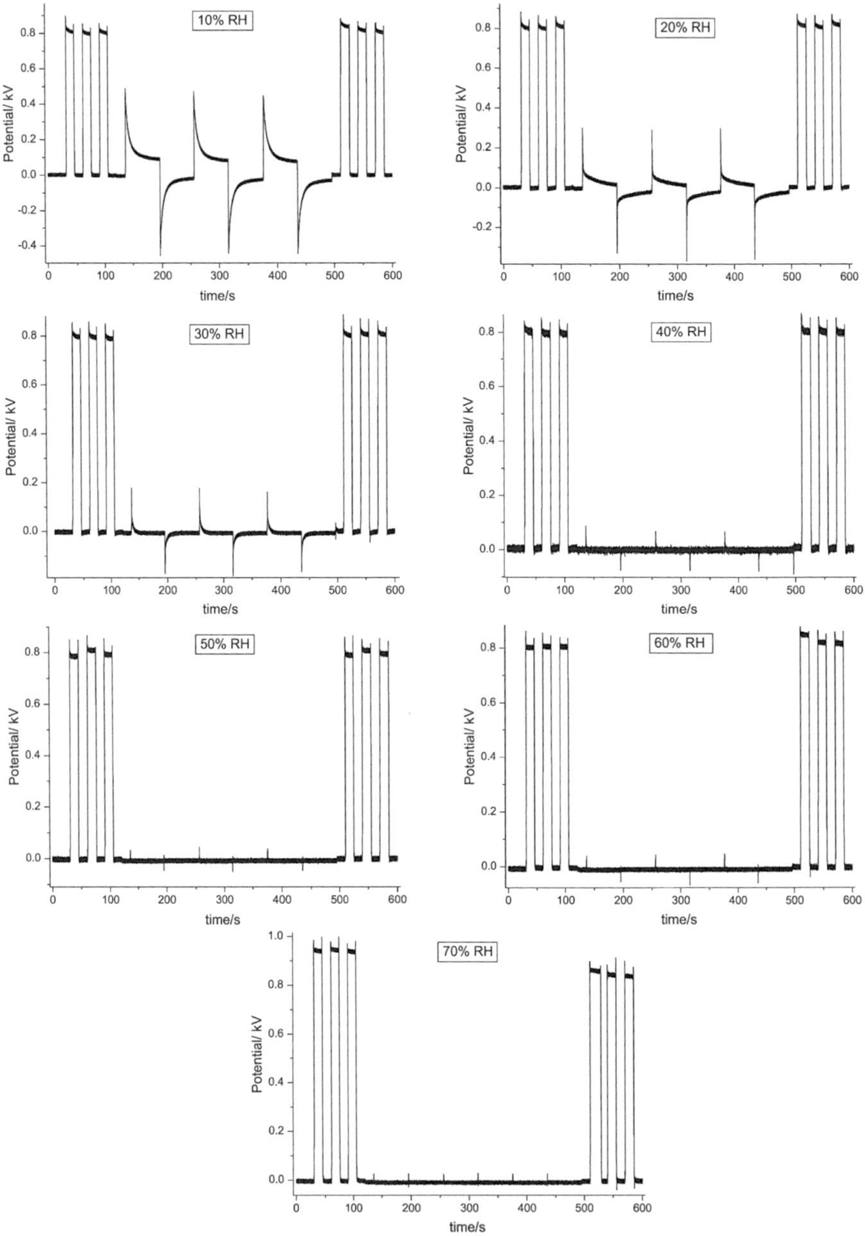

Fig. 6.2 Potential vs. time curves recorded for paper sheet under nitrogen, 10–70% relative humidity. Details are as in the legend of Fig. 6.1 Reprinted with permission from [7]

The effect of increasing relative humidity is shown in Fig. 6.2. Cellulose displays a significant shielding ability for DC and low-frequency AC fields, under RH equal or larger than 50%.

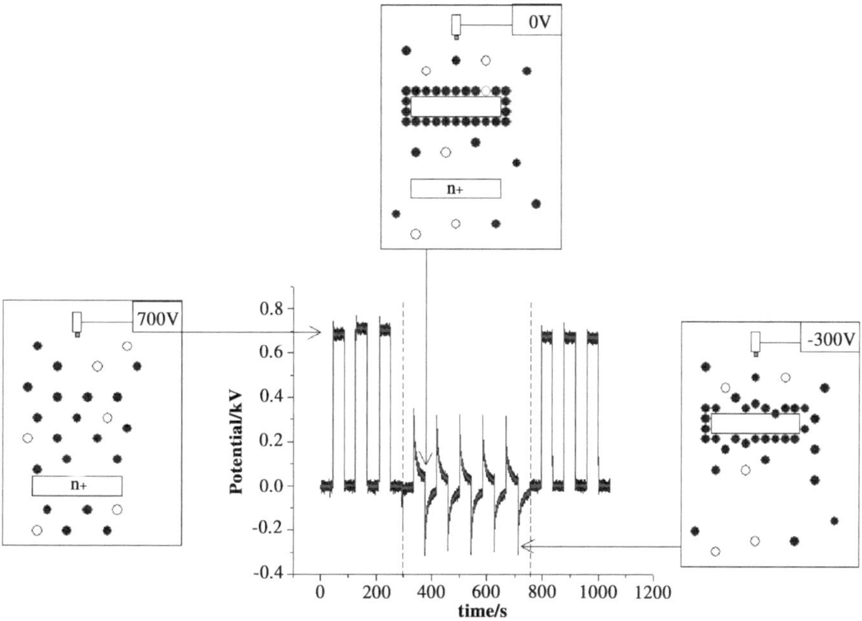

Fig. 6.3 Schematic representation of the mechanism proposed for the observed potential changes. *Left*: the poled inductor is introduced beneath the electrode that reads the potential created by all charges in the vicinity, both in the inductor as well as in the atmosphere. *Center*: the sample is introduced between the poled inductor and the electrode. The accumulation of charged water cluster ions in the sample surface leads to a decrease in the potential read by the electrode. *Right*: the poled inductor is withdrawn and the voltmeter reads a potential generated by the excess negative charge at the sample surface. *Black circles* are negative ion clusters and the *white circles* are positive charges. The potential vs. time plot in the center of the figure shows the points corresponding to the three states. Reprinted with permission from [7]

Figure 6.3 is a schematic representation of the main events in these experiments, showing how the accumulation of negative ions on cellulose under the positive potential of the charged acrylic sheet produces zero reading in the electrometer.

These results could not be explained using the usual induction models that are adequate for metals or semi-conductors, based on charge displacement across the insulator. They are also different from what would be expected if the shielding effect was solely based on water dipole orientation, because this is a much faster phenomenon with relaxation times in the microsecond range.

On the other hand, they are consistent with the hypothesis of sorption of atmospheric water on the insulator and its electrification, based on the effect of the applied *electric potential* on the *electrochemical potential* ($\mu_i = \mu_i^\circ + RT \ln a_i + zFV$). This produces excess concentrations of the ions formed by water adsorbed on cellulose, $H(H_2O)_n^+$ and $OH(H_2O)_n^-$.

This also helps to understand the nature of space charge associated with dielectrics in the presence of even minute amounts of water. Space charge is often mentioned but it is hardly identified with definite chemical species. These results led thus to a new model for electrostatic charging of dielectrics, based on excess $H(H_2O)_n^+$ and $OH(H_2O)_n^-$ ions formed and trapped within the solid surface and bulk.

Estimates of the concentration of excess charge from potential measurements yield very low charge densities. For instance, in an experiment of charge induction in paper sheets [7], carried out at 10% RH, 500 V were detected at the electrode probe. The calculated charge concentration is 1×10^{-2} unit charge/μm^3, only. Concentration-wise this is low but it is significant, as far as the resulting voltage is concerned.

6.3 Charging Metals with Atmospheric Humidity

In 2010, Ducati et al. [8] showed that the exposure of isolated metal samples to water vapor leads to the deposition of excess charge on the metal, measured using a Faraday cup.

A typical set of results is in Fig. 6.4, showing how periodic change in relative humidity produces the accumulation of negative charge on chrome-plated brass tubing.

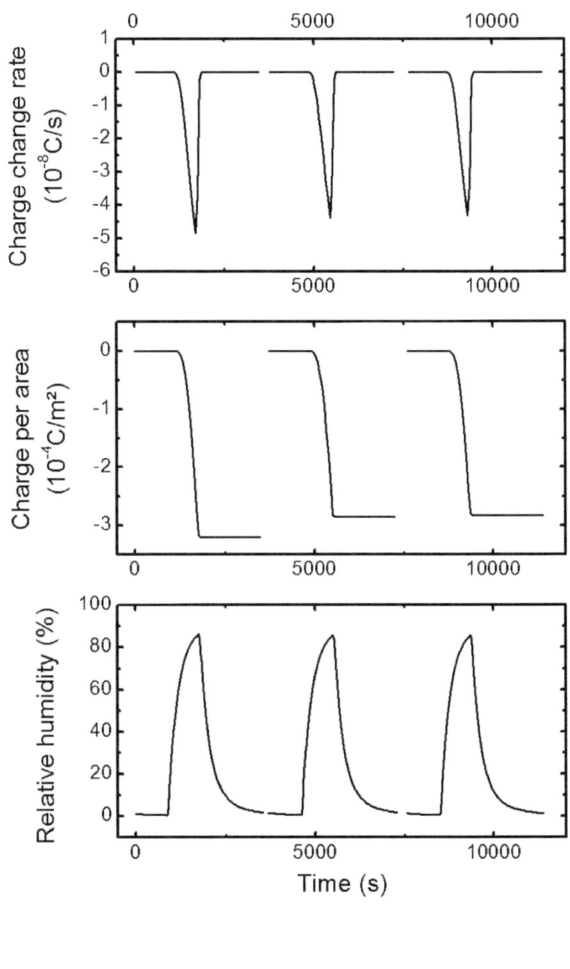

Fig. 6.4 RH, charge per area, and charge change rate vs. time plots for a chrome-plated-brass (CPB) tubing during dry-wet-dry N_2 cycles. Reprinted with permission from [8]

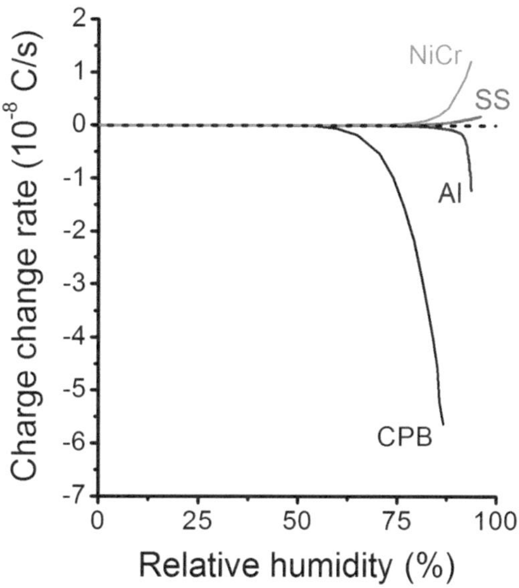

Fig. 6.5 Charge change rate dependence on increasing RH for NiCr, stainless steel (SS), aluminum (Al) and chrome-plated-brass (CPB). Rates for brass, copper, silicone-coated aluminum and stainless steel are negligible in the scale shown here. Reprinted with permission from [8]

Experiments like these were performed using many different metals and the accumulated results for aluminum, stainless steel, nichrome, brass, chrome-plated brass and copper are in Fig. 6.5. Copper and brass acquire little charge, nichrome and stainless steel become positive while aluminum and Cr-plated brass become negative.

CPB is not a usual laboratory material but it was important for the discovery of hygroelectricity, since it acquires higher charge than any other material tested so far. This kind of tubing was acquired by a student in the author's laboratory, who was in a hurry to build a Faraday cup and did not want to wait for delivery of the copper tubing. Instead, she found this tubing in a store selling building and home repair materials, and acquired it. If the student would build a more standard Faraday cup using copper metal, hygroelectricity would not have been discovered, probably.

Charging isolated metals within shielded and grounded containers without resorting to any source of electricity shows that the atmosphere can transfer charge to the metals. Thus, it is a charge reservoir for both positive and negative charge. Since the positive or negative current entering the otherwise electrically isolated metal is strongly dependent on the relative humidity, it is assigned to OH^- and H^+ ions transfer to gas–solid interfaces, producing net current. The electricity buildup dependence on humidity, or hygroelectricity, acts simultaneously but in opposition to the well-known charge dissipation due to the increase in surface conductance of solids under high humidity. Indeed, metal charging by adsorption of water vapor ions beats surface conductance.

The reason for obtaining positive or negative charge is assigned to specific adsorption on the metal surface or else, on the oxide layer coating the metal: if this adsorbs preferentially OH^- or H^+, it will acquire excess negative or positive charge. This is represented in Fig. 6.6.

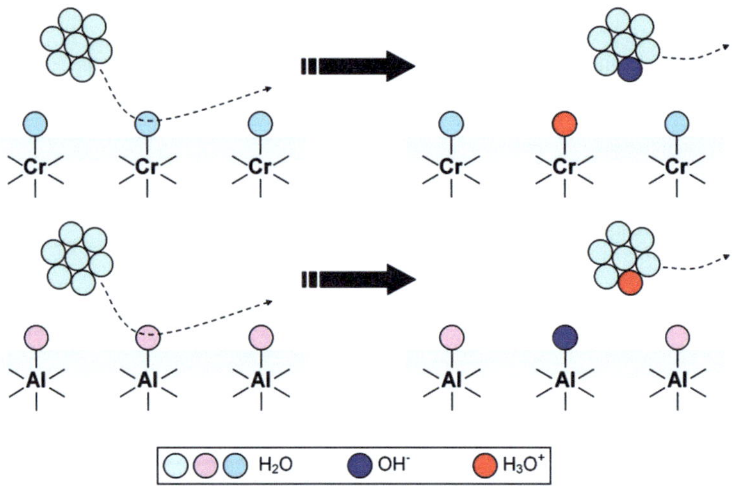

Fig. 6.6 Mechanism for charge transfer from the atmosphere to the metal surface. (*Top*) Formation of positive charge over a basic oxide. (*Bottom*) Formation of negative charge over an acidic oxide. Neutral water molecules are amphoteric, reacting differently with different oxides according to their acid–base properties. Reprinted with permission from [8]

This phenomenon holds the potential to produce power by scavenging electricity from the atmosphere. Hygroelectricity cells were built in the author's group, combining aluminum as the negative electrode using stainless steel and carbon as the positive electrode. This is an asymmetric capacitor, self-charging to 0.7 V spontaneously under 80% RH. Unfortunately, repeated charge/discharge cycles of the capacitor are accompanied by aluminum oxidation, which is highly undesirable. Consequently, practical use of hygroelectricity as a way to scavenge energy from the ambient still depends on finding more durable materials. Nevertheless, a group in Indonesia concluded that "... Indonesia has the potential to harness hygroelectricity to solve the energy crisis and environmental issues" [9].

6.4 The Effect of Humidity on Surface Charge Patterns

6.4.1 Water Vapor Adsorption and Desorption Modifies Charge Patterns

The discovery of the effect of changing relative humidity on the excess charge in metals and cellulose came soon after the independent discovery of charge patterns in almost any dielectric particles and films that were examined by scanning probe electric microscopies. For this reason, Kelvin micrographs were acquired from different materials under high and low humidity, starting with films of non-crystalline silica and aluminum phosphate fine particles [10], followed by many other solids.

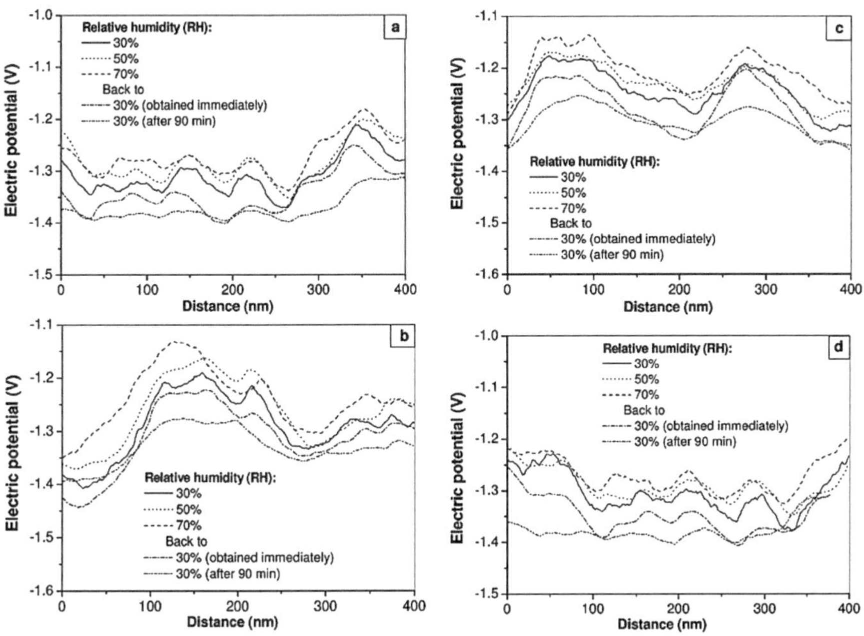

Fig. 6.7 Line-scans from the five consecutive KPM images from aluminum phosphate particles. Reprinted with permission from [10]

In every case that was examined, the electrostatic potential patterns change with the relative humidity, within an electrically shielded and grounded environment, as shown in Figs. 6.7 and 6.8, obtained for aluminum phosphate and Stöber silica nanoparticle films.

Moreover, the observed change is not uniform, across the samples. Potential adjacent to the particle surfaces is always negative and Fig. 6.9 shows potential gradients in excess of 10 MV/m, parallel to the film surfaces.

The examination of many other solids with known acid–base character [11] showed that the electrostatic potential at the surface of acidic solids becomes more negative under higher relative humidity (RH), while basic solids acquire a more positive potential at high RH, using the Kelvin method at the nano- and macroscales, see Table 6.1.

Images acquired in the Kelvin microscope under low and high humidity are in Fig. 6.10, together with the corresponding potential traces. Silica, iron oxide on iron and cellulose show parallel traces evidencing that the surface is uniform, concerning its acid–base character. On the other hand, aluminum oxide on metal is rather non-uniform. Moreover, changes on silica are highly reversible but on cellulose they are not.

Table 6.1 sums up the results obtained for nine solids. The average potential variations, ΔV for iron oxide on iron and calcium oxide are very large and the uniformity parameter $R = (\Delta V_m)2/(\Delta V)^2_m$ approaches unity for a few solids but it is only 0.4 for MgO, showing that this has a very non-uniform surface.

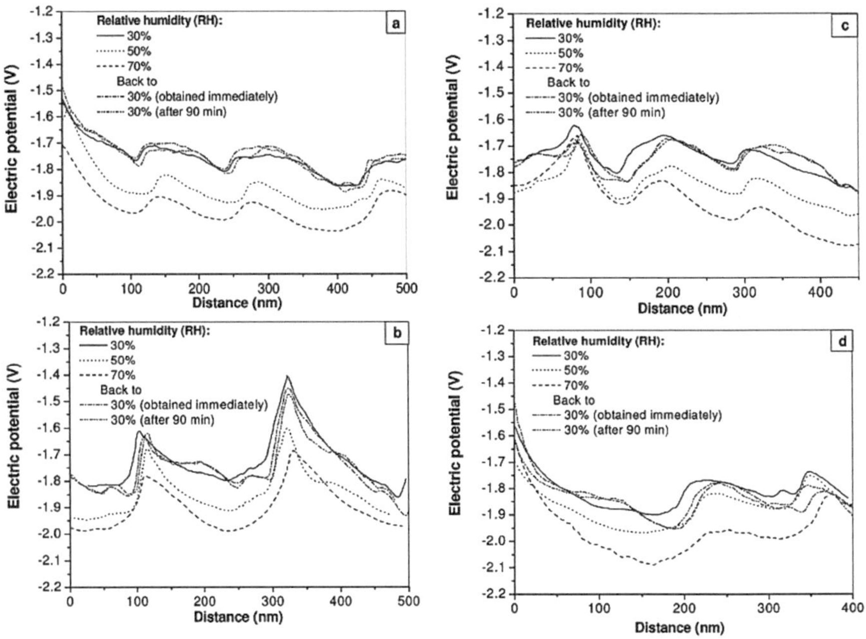

Fig. 6.8 Line-scans from the five consecutive KPM images from Stöber silica particles. Reprinted with permission from [10]

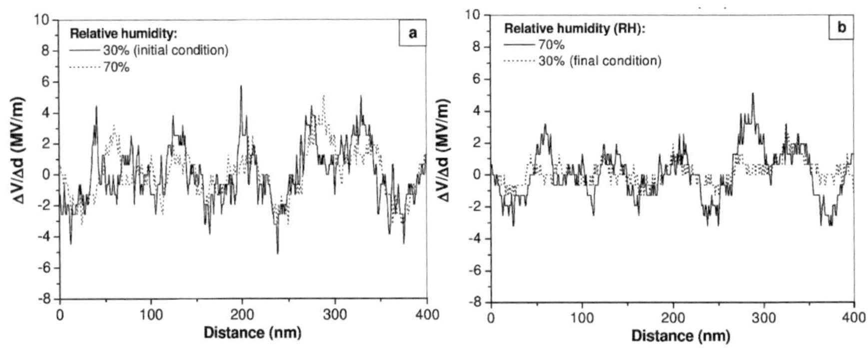

Fig. 6.9 Electric potential gradients calculated as a function of the sample position for Stöber silica particles after equilibrating for 90 min. (**a**) 30% RH (*solid line*, initial condition) and 70% RH (*dot line*). (**b**) 70% RH (*solid line*) and 30% RH (*dot line*, final condition). Reprinted with permission from [10]

This provided a basis for the selectivity of H^+ or OH^- adsorption from water vapor: acidic surfaces adsorb OH^- while basic surfaces adsorb H^+. Thus, the key factor for predicting the effect of changing humidity on KFM microscopy is the Brønsted acid or base character of the solid under examination. Kelvin microscopy

Table 6.1 Average electrostatic potential variation (ΔV_m) and the mean square $(\Delta V)^2_m$ for the analytical solids, following an increase in humidity (from 30% RH to 70% RH) [11]

Compound	ΔV_m	$(\Delta V_m)^2$	$R = (\Delta V_m)^2/(\Delta V)^2_m$	Assigment
Iron oxide	−0.464 ± 0.004	0.217 ± 0.004	0.99	Acid
Magnesium sulfate	−0.229 ± 0.008	0.063 ± 0.009	0.83	Acid[a]
Silica	−0.172 ± 0.015	0.033 ± 0.005	0.90	Acid[b]
Cellulose	−0.104 ± 0.007	0.011 ± 0.001	0.98	Acid
Aluminum oxide	−0.055 ± 0.016	0.004 ± 0.002	0.76	Acid[a]
Calcium oxide	+1.657 ± 0.115	2.811 ± 0.335	0.98	Base[b]
Magnesium oxide[c]	+0.195 ± 0.062	0.096 ± 0.021	0.40	Base[a]
Nickel oxide	+0.060 ± 0.017	0.005 ± 0.002	0.72	Base
Aluminum phosphate	+0.039 ± 0.007	0.002 ± 0.001	0.76	Base[b]

[a]Data taken from [12]
[b]Data taken from [8]
[c]30–50% RH only. This cannot be measured at 70% RH because of excessive water absorption

under variable humidity also provides information on the spatial distribution of acid–base sites, which is currently inaccessible to any other technique.

These results enlarged the scope of Kelvin force microscopy (KFM) that was then identified as a powerful, sensitive, and convenient technique for the detection and characterization of acid and base sites on solid surfaces, including adsorbents and catalysts. It shows many positive characteristics:

1. The only reagent used is water vapor. Thanks to its amphoteric behavior, water detects both acidic and basic sites;
2. The activity of the reagent water vapor is easily changed by changing the relative humidity;
3. Detection sensitivity is very large, especially when compared to widespread surface analytical techniques, since it is based on charge and potential measurements;
4. Uniformity (or not) of acid – base sites is easily detected, with a spatial resolution in the 10–20 nm range;
5. Standard noncontact atomic force microscopy (AFM) micrographs are acquired simultaneously with Kelvin images, revealing concurrent changes in the solid morphology, if any;
6. Kelvin microscopes are currently commercially available from many suppliers as attachments to AFM instruments.

Many perspectives are now open to this technique, since measurements can be done on different time scales in the range of many minutes and up, allowing the acquisition of kinetic information. Moreover, the spatial resolution can probably be improved in very smooth surfaces, such as those in well-defined single crystals, using high-resolution Kelvin microscopy [13].

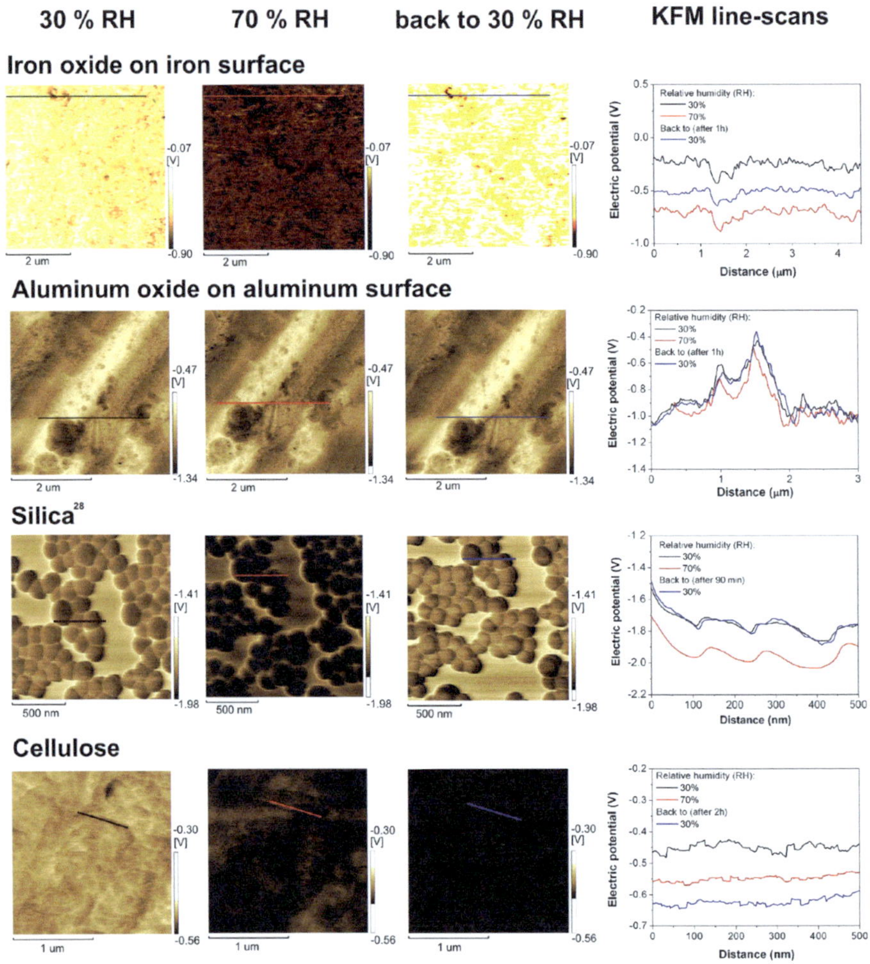

Fig. 6.10 KFM images and line-scans from the same area of iron oxide on metal, aluminum oxide on aluminum metal, silica and cellulose. Successive changes in the relative humidity (RH) were made prior to image acquisition, as indicated at the top of the figure. Darker areas are more negative than the brighter areas. The observed changes are at least partly reversible, except in the case of cellulose. Reprinted with permission from [11]

Many more additional parameters can be examined: sample temperature cycling effects, the presence of other gases [14] in the surrounding atmosphere (including hydrocarbons, alcohols, ammonia, pyridine, and CO_2), and the effect of pressure, within the limits defined by the available instrumentation.

To conclude, KFM that was initially designed to map electric potential holds an unmatched potential to advance knowledge on the acid – base behavior of solid surfaces, producing hitherto unavailable analytical information on important chemical systems.

6.4.2 Charge Build-Up on KFM Calibration Samples

Kelvin probe or Kelvin force microscopes (KFM) and other types of scanning probe instruments derived from the atomic force microscope have been playing a key role in the progress in discovering electrostatic patterns in surfaces in the last 20 years. This in turn led to radical revision of some widespread ideas that lacked strong experimental basis.

To make sure that the measurements made on a Kelvin microscope are meaningful, it requires calibrations as any other instrument. Calibration of x, y and z cartesian axis is done using standards that are verified by many other instruments, including the electron microscopes. To calibrate the electric potential measurements, a silicon wafer is oxidized forming a thin silica layer where a set of interdigitated electrodes made of evaporated gold is mounted, using microlithography techniques. The electrodes are connected to a precision DC voltage source and the potential on the electrodes is read using the Kelvin microscope set-up. During this calibration work, Rubia Gouveia noticed, in the author's laboratory, a pronounced effect of relative humidity shown in Fig. 6.11: the electric potential over silica is steady, under 10 or 30% but it changes pronouncedly with time, under 70%. Potential vs. time plots obtained during these experiments are shown in Fig. 6.12 [15].

The comparison of results obtained using the same protocol but under 10 or 70% relative humidity show marked differences and many intriguing points that were discussed in the paper. Most relevant to the present discussion is the pronounced difference between the potential vs. time plots for the silica surfaces, that was interpreted using the mechanism schematically depicted in Fig. 6.13. Electrode bias produces fast changes within times well-below 1 s followed by slow changes extending for many minutes.

Potential vs. distance plots were also acquired after the electrodes were short-circuited and grounded, at each RH. These are shown in Fig. 6.13, and they are also strongly dependent on the relative humidity. Line-scans measured at 70 and 50% RH show local potentials down to −1.2 V over silica and up to 0.3 V at the metal borders, forming regular, persistent patterns. The curves recorded under low RH also show deviations from zero but much smaller. This confirms that fixed charges are produced on silica while the electrodes are biased, and this is increasingly more pronounced at high RH.

All these observations are explained assuming the model presented in Fig. 6.14. Steep potential changes are observed following electrode connection to the power supply, followed by the formation of excess negative ion concentration in the surrounding atmosphere followed by the events described in the caption of Fig. 6.14.

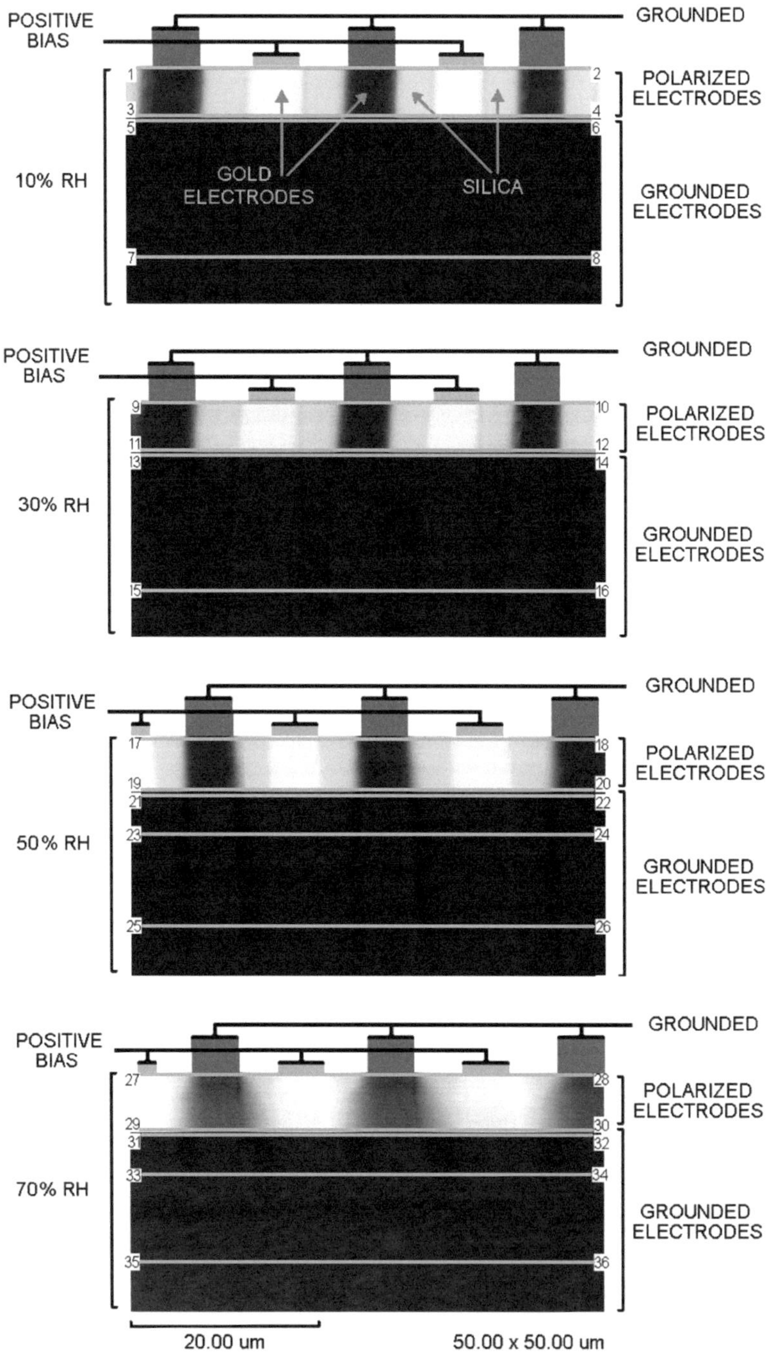

Fig. 6.11 KFM micrograph of a silica-on-wafer thin film partially covered with interdigitated electrodes. Successive changes in the state of electrode polarization and relative humidity of the surrounding atmosphere were made while the image was acquired, as indicated at the sides of the figures. Brighter areas are positive; dark areas are negative. Reprinted with permission from [15]

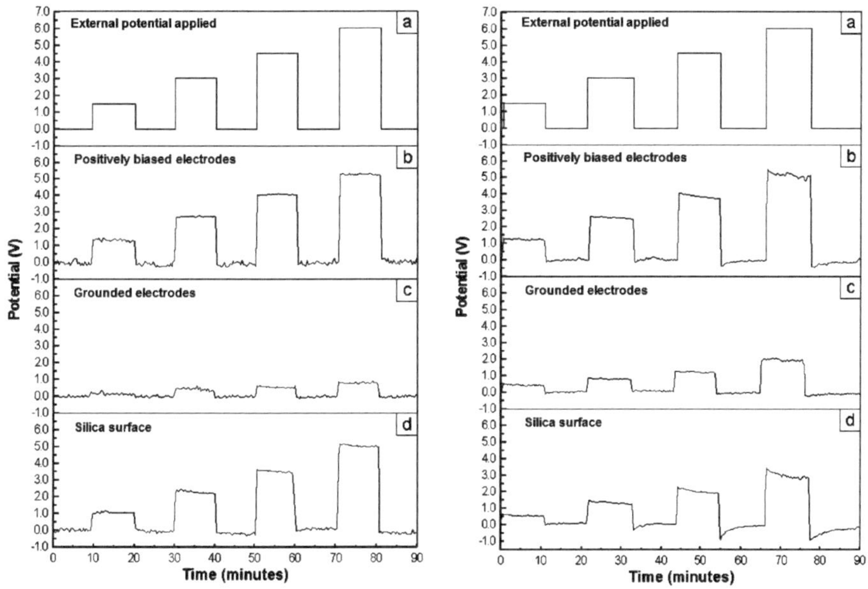

Fig. 6.12 (**a**) Potential difference applied to the metal electrodes. (**b**) Measured KFM potential over the biased electrode surface. (**c**) Measured potential over the grounded electrode. (**d**) Measured potential over the silica insulator in between the two electrodes. *Left:* 10% relatively humidity. *Right*: 70% relatively humidity. Reprinted with permission from [15]

6.4.3 Excess Charge Decay Through the Atmosphere

Water is usually seen as a passive agent for charge dissipation due to its low (but non-zero) electrical conductivity, following, for instance, the conclusions of Erwin Schrödinger's doctoral thesis [16]. It follows that charge elimination by surface conductance is the most often used mechanism for explaining the spontaneous decrease of charge under high humidity.

Saever [17] did detailed examination of this subject, in an attempt to explain the discrepancies reported in the literature, concluding that the surface resistivity of insulators is a measure of the amount of adsorbed water. Consequently, the relative humidity, temperature and the adsorption equilibrium of oxide pollutants and salts in the local atmosphere are all relevant variables responsible for the disparity in results obtained in different laboratories. It is important to keep in mind that surface water films are often discontinuous, due to dewetting phenomena [18, 19].

The role of the atmosphere as a source and sink of ions that was introduced in Sect. 6.2 contributes another path for charge transfer coupled to adsorption and desorption of H^+ and OH^- bound to water molecule clusters. Direct verification of the effectiveness of charge transfer to the atmosphere was obtained by following the discharge of corona- and tribocharged materials as a function of time and relative humidity. Quadruplicate samples were charged and allowed to discharge under

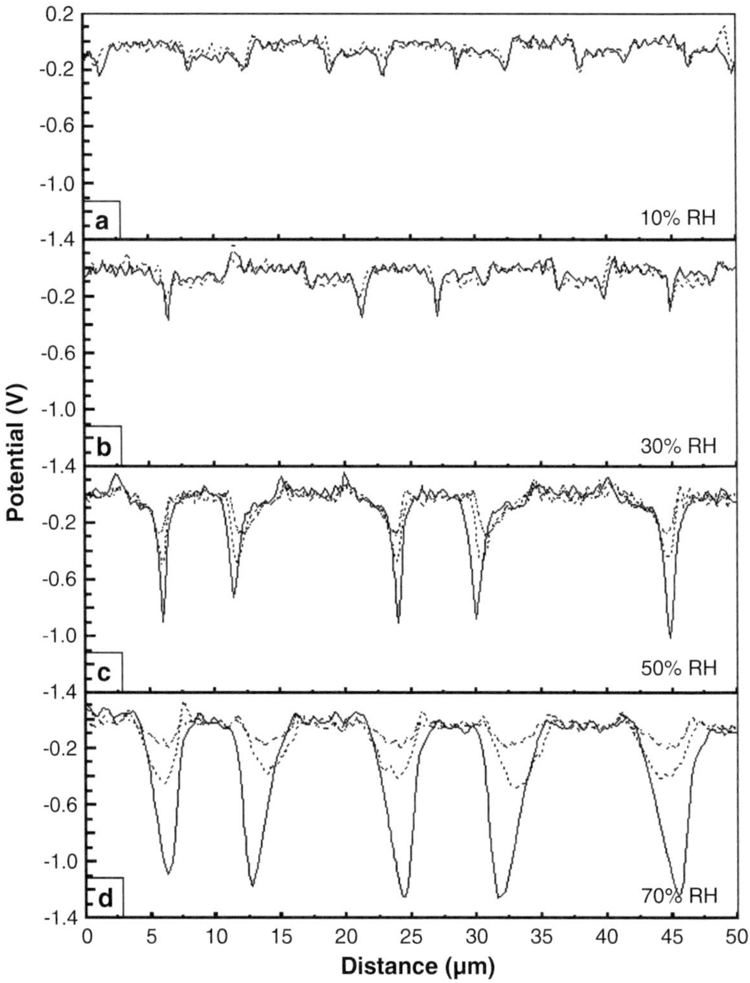

Fig. 6.13 Potential vs. distance along horizontal lines drawn in Fig. 6.11. *Solid line*: profiles acquired immediately after electrodes were grounded. *Dash line*: profiles acquired 10 min after electrodes grounding. *Dash dot line*: profiles acquired 30 min after electrodes grounding. (**a**) Profiles 5,6 (*solid line*) and 7,8 (*dash line*); (**b**) profiles 13,14 (*solid line*) and 15,16 (*dash line*); (**c**) profiles 21,22 (*solid line*), 23,24 (*dash line*) and 25,26 (*dash dot line*) and (**d**) profiles 31,32 (*solid line*) and 33,34 (*dash line*) and 35,36 (*dash dot line*), under 10, 30, 50, and 70% RH, respectively. Reprinted with permission from [15]

controlled relative humidity, while potential maps were recorded using a scanning Kelvin electrode, yielding potential half-lives.

Recorded curves for adjacent areas are uncorrelated, showing that charge surface motion is not the dominating mechanism for discharge. Relevant data are in Fig. 6.15. It shows half-lives of electric potential as a function of position for LDPE samples previously charged with positive and negative corona [20]. There are

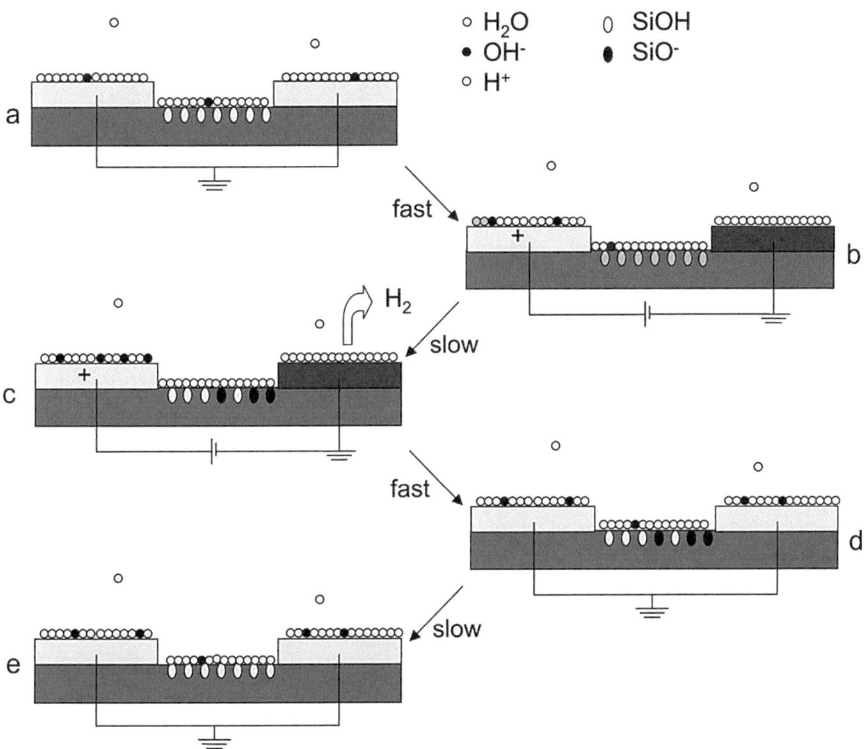

Fig. 6.14 Schematic representation of the model for the behavior of water molecule in the electrodes and silica surface, as well as of the silanol groups on silica. (**a**) In the initial state, the water film is neutral. At the gas–solid interfaces, (**b**) when an electrode set is biased, surface ions migrate to electrodes carrying potential of opposite signal; (**c**) silanol groups are slowly converted to silicate while H^+ ions are discharged at the grounded electrode; (**d**) when the electrodes are all grounded, ions at the surface film migrate reforming a neutral water layer; (**e**) silicate groups bind H^+ ions from water layer; (**e**) silicate groups bind H^+ ions from water and they are thus neutralized. Reprinted with permission from [15]

significant differences between potential half-lives within any given sample, they are shorter for negative samples and highly dependent on the relative humidity.

An unexpected finding was that all the potential vs. time curves converge to low negative values (−4.6 ± 0.7 V). This non-dissipated potential was then called *equilibrium potential* [20] and it was observed for other hydrophobic polymers.

The following mechanism is consistent with these results:

1. OH^- ions from the vapor bind to surface sites S_{OH} on LDPE with rate $v_{dp} = [OH^-]_v [S_{OH}]$.
2. H^+ ions from the vapor bind to surface sites S_{OH} on LDPE with rate $v_{dn} = [H^+]_v [S_H]$.

Since $v_{dp} > v_{dn}$, $K = v_{dp}/v_{dn} = [S_{OH}]/[S_H] > 1$, provided the $[OH^-]_v/[H^+]_v$ ratio is approximately equal to 1. This assumption is consistent with the symmetry for formation of positive and negative ions from water that was observed in other sections in this chapter.

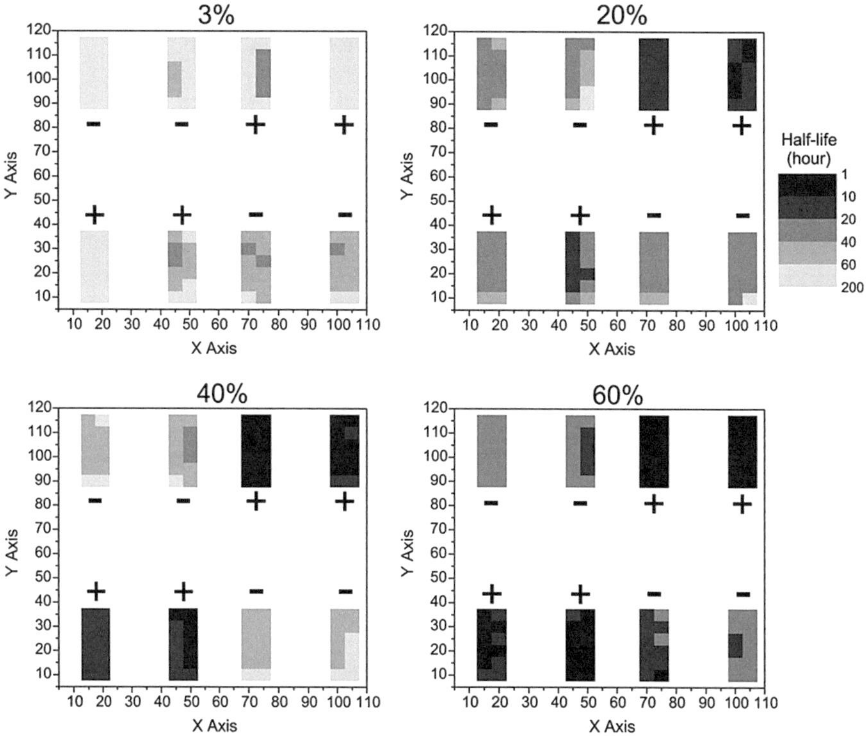

Fig. 6.15 Maps showing the half-lives of electric potential decay as a function of the position on LDPE pieces under variable relative humidity. Reprinted with permission from [20]

This mechanism explains all the experimental observations made in this experiment, placing it within a simple conceptual framework that is familiar even to high-school students. Moreover, it is consistent with the idea of specific OH^- adsorption on hydrophobic surfaces [21, 22].

The importance of surface conduction as the prevalent mechanism for discharge is also dismissed by experimental results on the discharge of tribocharged materials. The successive potential maps recorded on rubbed PTFE (see Fig. 6.16) show faster decay of positive pixels and no indication of charge neutralization at the separation between positive and negative areas [23].

6.5 Flow Electrification: The Position of Water in the Triboelectric Series

Liquid charging during flow within a pipe is well known as *flow electrification (FE)* that is a known oil pipeline hazard. It was also recently observed for many other liquids, including water. For instance, ultrapure water used in the fabrication of semiconductor chips can cause electrostatic discharges in electronic components

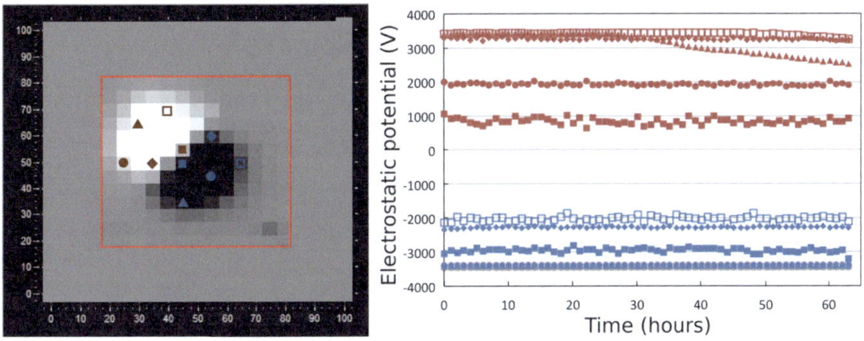

Fig. 6.16 Electrostatic potential variation on PTFE tribocharged with PE foam. (*Left*) Potential map of PTFE abraded with PE foam. (*Right*) Electrostatic potential of the selected pixels measured during 63 h at 60% relative humidity. Reprinted with permission from [23]

[24]. Other reports show that pure water flowing through hydrophobic materials such as polytetrafluorethylene (PTFE) becomes positively charged [25, 26]. However, there is only limited information on flow electrification of water with different surfaces [27].

A simple and robust apparatus provided reproducible results on excess electric charge acquired by water flowing through different materials [28]. The experimental setup is in Fig. 6.17a and a summary of results is presented in Fig. 6.17b. Water acquires a net positive charge against all solids tested and the charge magnitude follows most triboelectric series [29]. Acquisition of excess positive charge in water may be understood as a case of electrosmosis that is explained considering the preferential adsorption of OH^- ions at the solid surfaces. This produces an electric double layer with negative ions trapped at the Stern layer while the positive H^+ ions in the diffuse layer move with bulk water. This is represented in Fig. 6.17c.

This may also be related to the exclusion-zone formed on the vicinity of various hydrophilic surfaces [30, 31] that displays a negative net charge.

A recent report shows that water collected from various different places and placed within common containers (e.g. polyallomer centrifuge tube) is always negatively charged [32], in apparent disagreement with the results in Fig. 6.17. However, these measurements were made using a Faraday cup containing the whole system, water plus its recipient, not just water itself.

6.6 Water Dropping from a Biased Needle

Water with excess charge is obtained by connecting to a biased electrode and it can be stored in a Faraday cup [33].

Water dropping from an electrically biased needle acquires excess charge as shown in Fig. 6.18a. The charge sign is the same as the sign of the needle potential and it modifies the water surface tension, decreasing it pronouncedly, as in Fig. 6.18b.

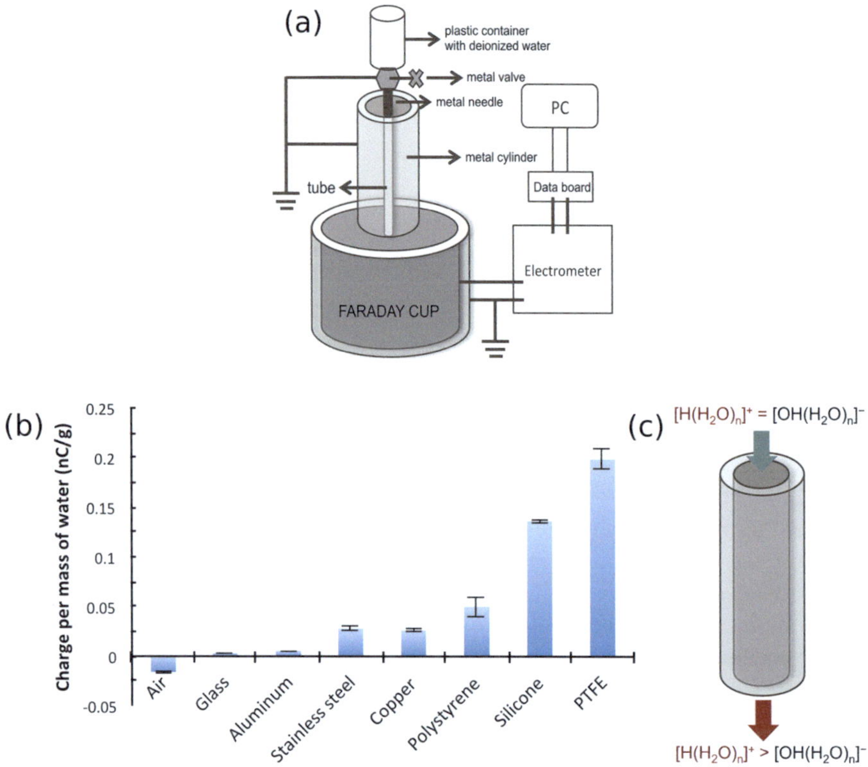

Fig. 6.17 (**a**) Experimental setup used to measure electric charge of water acquired by flowing through different materials; (**b**) Charge per unit weight of water following contact with different materials; (**c**) Schematic charging mechanism: neutral water enters the tubing where OH^- ions are adsorbed while H^+ ions remain within water. Reprinted with permission from [28]

This was verified by using the drop shape and drop volume/weight measurement techniques. The drops are distorted as the potential increases until the drops stretch into streaks of electrified liquid, at $V > 9$ kV, showing that electrostatic repulsion overcomes surface tension. Under high tension, sessile water drops resting on top of biased needles undergo Columbic explosion, sending smaller droplets upward.

Measuring charge, potential and surface tension of the electrified liquid provides all the information usually obtained from electrocapillarity experiments. This has been a limited topic of study due to the scarcity of conducting non-reactive liquids that made electrocapillarity virtually exclusive to the mercury electrodes that are not popular due to toxicity and environmental problems. However, the experimental protocols used by Santos et al. [33] are adaptable to most common liquids at room temperature and this brings feasibility to the study of electrocapillarity. One important outcome of these experiments is the demonstration that potential drop at the water–air interface is a few tens of volts only, when the water drop is biased in the kV region. This is the first estimate of potential drop within the electric double layer

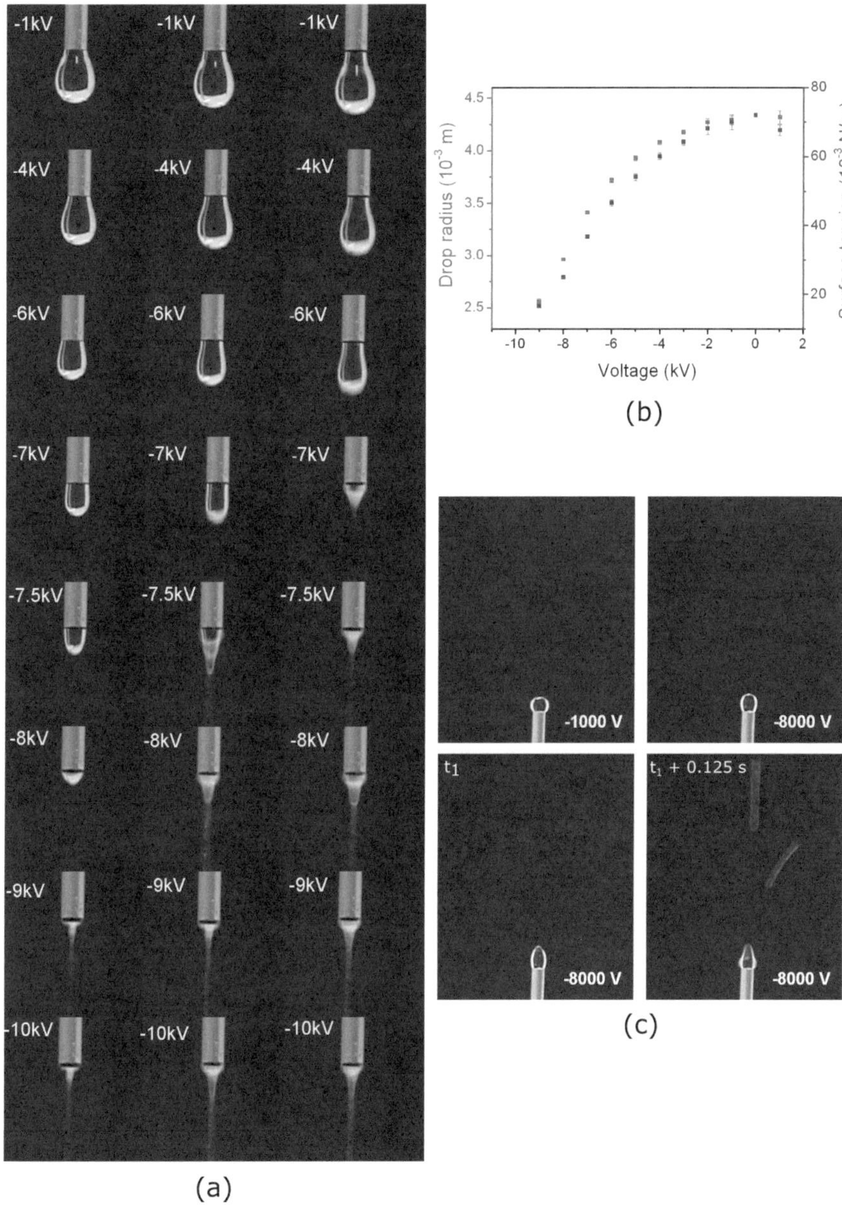

Fig. 6.18 (**a**) Water exiting the tip of a biased needle, showing Taylor cone formation, drop elongation, and Coulombic explosion at high V. Water flow: 64.1 mL/h. Distance from needle tip to the nearest grounded surface was ~10 cm. Pictures were chosen from a recorded video to represent interesting features, and time interval between them is not uniform; (**b**) Drop radius and calculated surface tension as a function of needle electric potential. Error bars are always plotted, but in some points they are smaller than the symbol. Minimum distance between needle tip and surrounding grounded surfaces was 1.1 m, to minimize electric field strength; (**c**) Coulombic explosion of a water sessile drop when the supporting needle is biased to −8000 V. Notice the faint streak on top of the photo. Parts of this figure are reprinted with permission from [33]

at a gas/liquid interface, ever. On the other hand, water density and viscosity are unaltered, showing that charges are accumulated at interfaces only and do not affect bulk properties.

6.7 Spontaneous Electric-Bipolar Nature of Aerosols

Aerosols are widely found on the Earth's atmosphere and have paramount importance in atmospheric phenomena [34], including a large number of chemicals discharged in the atmosphere by natural and anthropic phenomena. Aqueous aerosols are particularly relevant for cloud formation, stability and rain precipitation, while clouds are also important as precursors of atmospheric aerosols formed by other chemical substances [35, 36]. Many authors in this area relate atmospheric electricity to liquid water or ice particles and assign its formation to ice particle breakdown within storm clouds, while others give great importance to air ionization due to radiation.

Aerosol formation, properties and stability are often related to important practical problems in industrial, energy and health contexts. For this reason, there is excellent information on charge in aerosols [37] but this is not shown in the literature in atmospheric chemistry, as frequently as in the literature on colloidal sols, where zeta potentials or particle charge data appear in nearly every paper.

A robust and reproducible method to generate a current of aerosol is a nebulizer used for domestic inhalation. Charge in aerosol is detected by targeting the flowing aerosol to the interior of a Faraday cup while simultaneously measuring the electrostatic potential of the nebulizer with residual water, using a macroscopic Kelvin electrode [38], as shown in Fig. 6.19a. Moreover, aerosol electrophoresis is done by passing the aerosol in between two parallel copper disks connected to a high voltage power supply, as in Fig. 6.19b. These two procedures do not give detailed information on particle size and charge distribution but they yield information on particles as they leave the aerosol source, with minimal particle manipulation [39].

When aerosol from deionized water stemming out of a nebulizer passes through the inner electrode of the Faraday cup it produces initially highly variable positive current (Fig. 6.20a), changing to a time series of sharp positive and negative current peaks. When the aerosol is made using an aqueous NaCl solution, a similar result is obtained but the initial current is negative (Fig. 6.20b). The positive and negative peaks show that the aerosol is bipolar, this means, it contain both positive and negative droplets.

Observation of aerosol motion within an electric field confirms the bipolar charging behavior: when the aerosol enters the inter-electrode volume, it divides into convective streaks that migrate separately toward each electrode. Some suggestive frames are shown in Fig. 6.21. In control runs, aerosol flows undistorted in between grounded electrodes. Aerosol from deionized water deviates largely toward the negative electrode but a significant amount migrates to the positive electrode. Aerosol from NaCl solution also deviates, but the bulkier stream moves toward the positive pole.

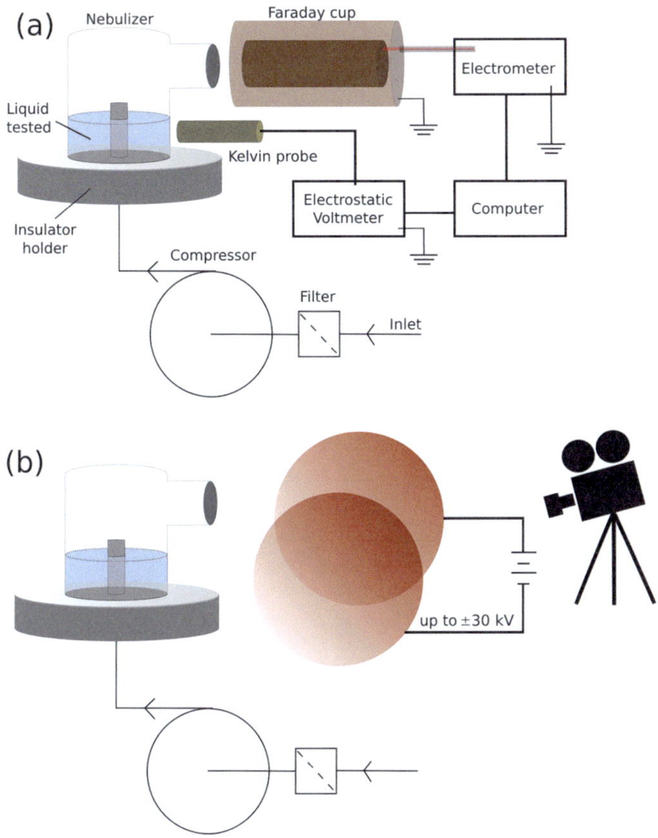

Fig. 6.19 Setups to measure aerosol charging and electrophoresis. Reprinted with permission from [39]

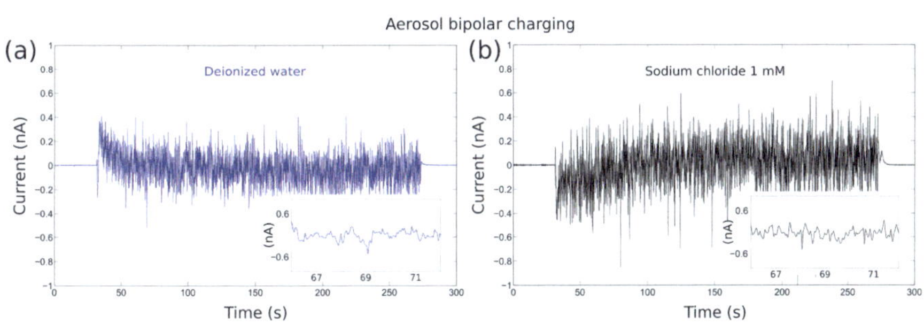

Fig. 6.20 Electric current between the electrodes of a Faraday cup, produced by passing aerosol from (a) deionized water and (b) sodium chloride solution (1 mmol L^{-1}) measured at a high-speed acquisition rate (800 readings/s). Reprinted with permission from [39]

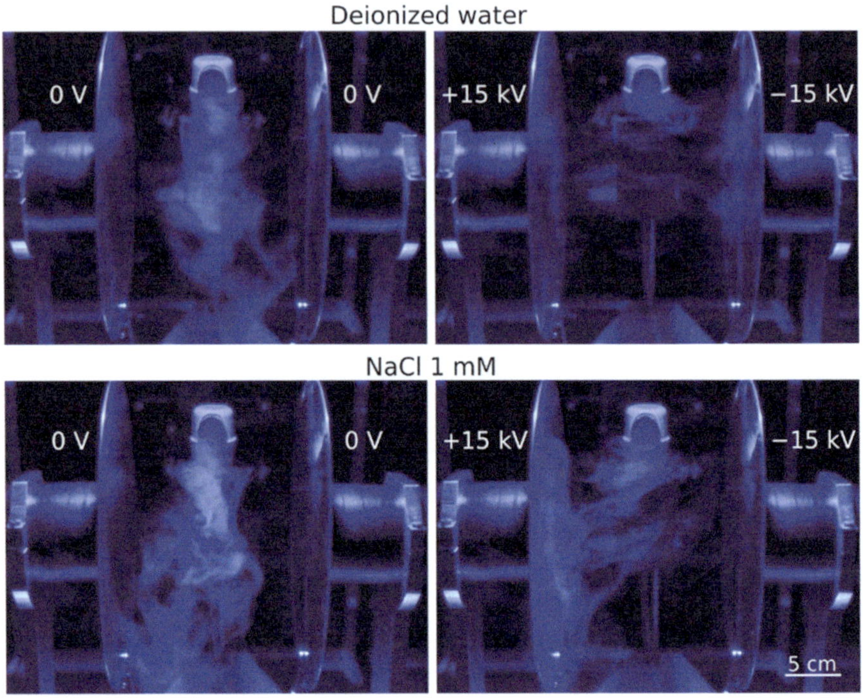

Fig. 6.21 Aerosol electrophoresis. Liquid is deposited on both electrodes, revealing that both positive and negatively charged droplets coexist in the aerosol. Reprinted with permission from [39]

These results show that liquid fragmentation by dropping, splashing and other frequent phenomena introduces electrostatic charge into the environment, where it is absorbed or adsorbed in the surrounding objects. This shows that the atmosphere is always filled with ions derived from water, salt and other common substances. For this reason, the often-quoted ionization caused by high-energy particles crossing the atmosphere is not essential for the appearance of atmospheric electricity [38, 39], even though it certainly makes a contribution.

The new understanding on charge bipolarity in aerosols may contribute to progress in scavenging electrical energy from the atmosphere and to increase safety while handling liquids in industrial environments. Finally, enlarging current knowledge on aerosol charging phenomena and mechanisms will probably contribute to a better understanding of atmospheric charge formation and stability.

6.8 Conclusion and Prospects

Water plays a dual role in electrostatic phenomena. It contributes to discharging by surface and bulk conduction thanks to the mobility of its ions, $H(H_2O)_n^+$ and $OH(H_2O)_n^-$ but it also plays the opposite role, acting an electrifying agent through different processes: adsorption of vapor on metals and insulators and partition during phase transition. Given its ability to partition and transfer ionic charge, water can also store charge reaching high potential. This has not yet been exploited for charge storage but the possibility is obvious with the perspective of achieving high energy density.

The electrification processes cover quite different potential ranges, from a few hundreds of millivolts to thousands of volts. Most relevant, charge transfer associated with mass transfer has a unique feature: it depends on the geometrical distribution of charge. This means, if a set of particles is charged to a given potential and then particles are brought to close proximity while maintaining their individual charges, the potential reached by the particle assembly will largely exceed the charging potential. This is at the core of functioning of the Pelletron high voltage machines but it is probably also related to the creation of hazardous conditions in handling powders and liquids.

References

1. Agam N, Berliner PR (2006) Dew formation and water vapor adsorption in semi-arid environments—a review. J Arid Environ 64:572–590
2. Ludwig R (2001) Water: from clusters to the bulk. Angew Chem Int Ed 40:1808–1827
3. Hagymassy J, Brunauer S et al (1969) Pore structure analysis by water vapor adsorption. J Colloid Interf Sci 29:485–491
4. Hutson ND, Yang RT (1997) Theoretical basis for the Dubinin-Radushkevitch (D-R) adsorption isotherm equation. Adsorption 3:189–195
5. Solomon I, Ribeiro AM et al (2013) Adsorption equilibrium of water vapor on activated carbon and alumina and carbon and alumina impregnated with hygroscopic salt. Turk J Chem 37:358–365
6. Homma Y, Chiashi S et al (2013) Photoluminescence measurements and molecular dynamics simulations of water adsorption on the hydrophobic surface of a carbon nanotube in water vapor. Phys Rev Lett 110:157402
7. Soares LC, Bertazzo S et al (2008) A new mechanism for the electrostatic charge build-up and dissipation in dielectrics. J Brazil Chem Soc 19(2):277–286
8. Ducati TRD, Simões LH et al (2010) Charge partitioning at gas-solid interfaces: humidity causes electricity buildup on metals. Lagmuir 26(17):13763–13766
9. Danaa CDP, Banazakia AS et al (2015) Humidity potential as hygroelectric power in Indonesia: opportunities and challenges. In: The 4th Indonesia EBTKE-ConEx, Jakarta, pp 127–130
10. Gouveia RF, Galembeck F (2009) Electrostatic charging of hydrophilic particles due to water adsorption. J Am Chem Soc 131:11381–11386
11. Gouveia RF, Bernardes JS et al (2012) Acid-base site detection and mapping on solid surfaces by Kelvin force microscopy (KFM). Anal Chem 84:10191–10198
12. Tanabe K, Misono M et al (eds) (1989) New solid acids and bases: their catalytic properties. In: Studies in surface science and catalysis, vol 51. Elsevier Science, New York
13. Barth C, Foster AS et al (2011) Recent trends in surface characterization and chemistry with high-resolution scanning force methods. Adv Mater 23:477–501
14. Janzen MC, Ponder JB et al (2006) Colorimetric sensor arrays for volatile organic compounds. Anal Chem 78:3591–3600
15. Gouveia RF, Costa CAR et al (2008) Water vapor adsorption effect on silica surface electrostatic patterning. J Phys Chem C 112:17193–17199

16. Schrödinger E (1910) Über die Leitung der Elektrizität auf der Oberfläche von Isolatoren an feuchter Luft. Ph.D. thesis, University of Wien
17. Seaver AE (2005) Surface resistivity of uncoated insulators. J Electrocardiol 63:203–222
18. Lee LT, Leite CAP et al (2004) Controlled nanoparticle assembly by dewetting of charged polymer solutions. Langmuir 20:4430
19. Bernardes JS, Rezende CA et al (2010) Morphology and self-arraying of SDS and DTAB dried on mica surface. Langmuir 26:7824–7832
20. Burgo TAL, Rezende CA et al (2011) Electric potential decay on polyethylene: role of atmospheric water on electric charge build-up and dissipation. J Electrostat 69:401–409
21. Healy TW, Fuerstenau DW (2007) The isoelectric point/point-of zero-charge of interfaces formed by aqueous solutions and nonpolar solids, liquids, and gases. J Colloid Interf Sci 309:183–188
22. Kudin KN, Car R (2008) Why are water-hydrophobic interfaces charged? J Am Chem Soc 130:3915–3919
23. Burgo TAL, Ducati TRD et al (2012) Triboelectricity: macroscopic charge patterns formed by self-arraying ions on polymer surfaces. Langmuir 28:7407–7416
24. Yatsuzuka K, Mizuno Y et al (1994) Electrification phenomena of pure water droplets dripping and sliding on a polymer surface. J Electrostat 32:157–171
25. Yatsuzuka K, Higashiyama Y et al (1996) Electrification of polymer surface caused by sliding ultrapure water. IEEE Trans Ind 32:825–831
26. Ravelo B, Duval F et al (2011) Demonstration of the triboelectricity effect by the flow of liquid water in the insulating pipe. J Electrostat 69:473–478
27. Paillat T, Moreau E et al (2000) Streaming electrification of a dielectric liquid through a glass capillary. In: Industry Applications Conference, 2000, Conference Record of the 2000 IEEE, vol 2, pp 743–748
28. Burgo TAL, Galembeck F et al (2016) Where is water in the triboelectric series? J Electrostat 80:30–33
29. Diaz AF, Felix-Navarro RM (2004) A semi-quantitative tribo-electric series for polymericmat erials:theinfluenceofchemical structure and properties. J Electrostat 62:277–290
30. Zheng JM, Chin W et al (2006) Surfaces and interfacial water: evidence that hydrophilic surfaces have long-range impact. Adv Colloid Interf Sci 127:19–27
31. Das R, Pollack GH (2013) Charge-based forces at the Nafion-water interface. Langmuir 29:2651–2658
32. Amin MS, Peterson TF Jr et al (2006) Advanced Faraday cage measurements of charge and open-circuit voltage using water dielectrics. J Electrostat 64:424–430
33. Santos LP, Ducati TRD et al (2011) Water with excess electric charge. J Phys Chem 115:11226–11232
34. Kaufman YJ, Tanré D et al (2002) A satellite view of aerosols in the climate system. Nature 419:215–223
35. Hirsikko A, Niemien T et al (2011) Atmospheric ions and nucleation: a review of observations. Atmos Chem Phys 11:767–798
36. Zhang R, Khalizov A et al (2012) Nucleation and growth of nanoparticles in the atmosphere. Chem Rev 112:1957–2011
37. Wong J, Knok PCL et al (2016) Bipolar electrostatic charge and mass distributions of powder aerosols—effects of inhaler design and inhaler material. J Aerosol Sci 95:104–117
38. Burgo TAL, Galembeck F (2016) Electrified water: liquid, vapor and aerosol. J Brazil Chem Soc 27:229–238
39. Burgo TAL, Galembeck F (2015) On the spontaneous electric-bipolar nature of aerosols formed by mechanical disruption of liquids. Colloids Interf Sci Commun 7:7–11

Chapter 7
Excess Charge in Solids: Electrets

Contents

7.1 Charge in Surfaces and in Bulk Solids

A frequent statement on charged materials is that "electrostatic charge is located at the surface". This is a logical outcome of the repulsion among charged particles, either positive or negative and it is observed in metals, in liquids (as shown in Chap. 6) and whenever electrons or ions are more or less freely mobile to migrate, driven to minimum Gibbs energy state.

F. Galembeck, T.A.L. Burgo, *Chemical Electrostatics*,
DOI 10.1007/978-3-319-52374-3_7

However, when charge excess is formed within some phase where charge transfer is hindered due to low mobility of the charge carriers, it remains trapped for as long as any charge suppressing phenomena are prevented.

Moreover, Chap. 6 in this book shows that the atmosphere is a charge reservoir due to charge transfer concurrent with water vapor adsorption–desorption due to H^+ or OH^- partition during adsorption. That contributes positive or negative charge to the surface and thus to the overall solid, depending on the properties of the solid surface. The mobility of water vapor in the atmosphere is very high but the rates of water adsorption-desorption are not. Consequently, charge build-up and dissipation through water exchange with the atmosphere is dependent on the nature and properties of the solid surface.

Solid surfaces are by themselves a complex topic and real surfaces show a huge variability that is not often acknowledged. Some aspects of solid surfaces relevant to electrostatic behavior are presented in the next section.

7.2 Relevant Features of Solid Surfaces

Material surfaces display some common features whose knowledge is essential to understand some problems found in electrostatics. For instance, it is quite common that the chemical composition of the solid surface differs from its interior, pronouncedly. Two examples are window glass and low-density polyethylene (LDPE) film used for packaging.

7.2.1 Glass and Other Hydrophilic Surfaces

Window glass is a non-crystalline sodium and calcium silicate, containing variable amounts of other components, especially magnesium and aluminum. It is produced and processed at high temperatures that remove even strongly bound water that is normally found in its precursors: sand, lime and sodium carbonate. Moreover, the ions in glass do not have significant mobility, below the glass transition temperature: diffusion coefficients and electrophoretic mobility are negligible. For these reasons, silicate glasses are strongly insulating and they are widely used to make insulators used in power transmission lines.

However, glass surfaces are quite different: they adsorb water from the surrounding atmosphere, including rain and aerosol. Water adsorption increases with the relative humidity but it is observed even under low humidity. The siloxane bonds that abound in the glass bulk are hydrated, forming the weakly acidic silanol groups, akin to silicic acid. Sodium and calcium ions can also hydrate, acquiring increasing mobility to the point of dissolving in the surrounding water, leaving behind a negative silicate surface.

Silica and silicate surfaces adsorb water, even under low relative humidity (RH). Silicate glass is highly variable but there are relevant data on well-defined silica

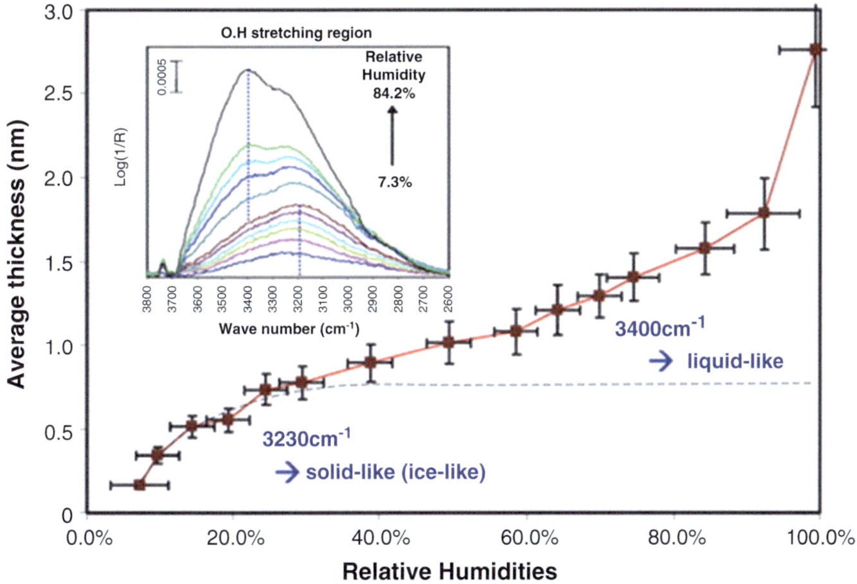

Fig. 7.1 Adsorption isotherms for water adsorption on silicon dioxide and complementary ATR-IR spectrum of O–H stretching region (inset): *dotted line* in the adsorption isotherm indicates thickness of adsorbed water in solid-like state as indicated by IR signal at 3230 cm^{-1}. Reprinted with permission from [1]

glass surfaces: average thickness on silica glass under 20% RH corresponds to a few water molecule layers that are easily detected using infrared spectra, growing to 1 nm thickness at 60% RH [1]. Judging from the IR spectra, water mobility at 20% RH is very low, justifying the "solid-like" designation often used in the literature. Under 60% RH the adsorbed water is a "liquid-like" layer, as shown in Fig. 7.1. Its thickness allows it to form a conductive electric double-layer with mobile ions. For these reasons, silicate glass surface is expected to show a completely different behavior from bulk glass, concerning electric conductivity.

The peculiar surface properties of glass surfaces contribute an interesting dependence of its electric behavior on the relative humidity. This is evidenced in Fig. 7.2 that shows electrostatic potential variation across a glass surface fitted with two parallel electrodes, under different values of the relative humidity. At 3% RH the electrostatic potential varies linearly on the region comprising the electrodes, as expected for a capacitor formed with two parallel electrodes. However, potential in the interelectrode space changes pronouncedly under higher relative humidity, becoming negative. When the electrodes are grounded, potential across the glass surface is nil, within experimental error. This result is analogous to a similar experiment that was done at a microscopic scale [2].

The observations made on glass cannot be simply extended to other hydrophilic solids and the consequences of the presence of adsorbed water also vary: surface water strengthens silica glass toward abrasion, but not copper oxide on copper.

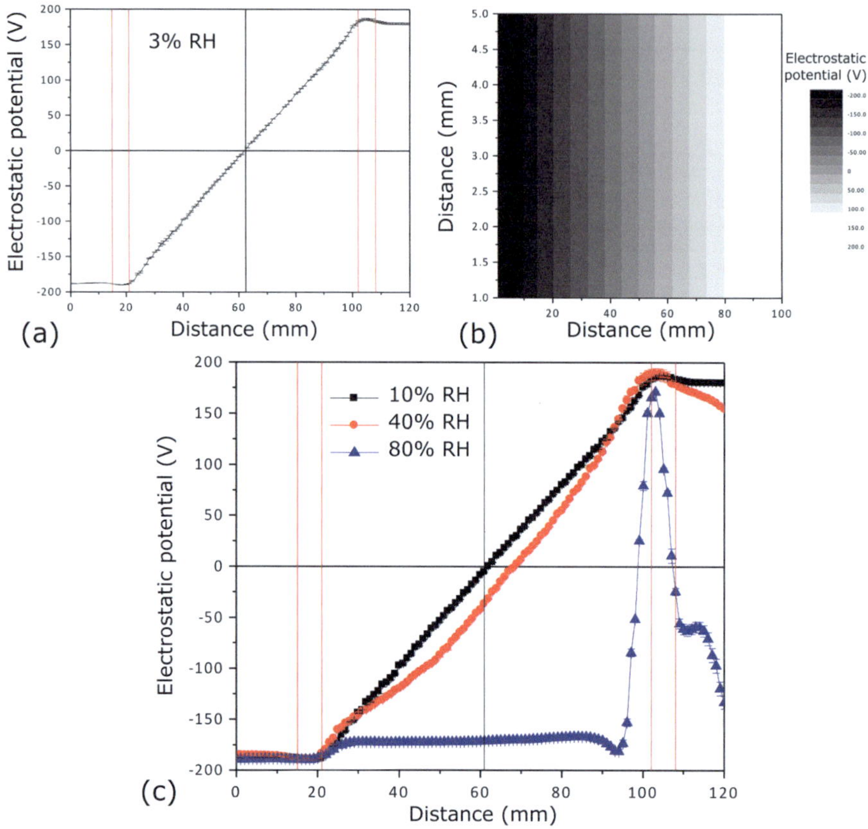

Fig. 7.2 Electrostatic potential of glass biased under ±184 V. In (**a**) relative humidity was kept at 3%. The electrostatic potential map (**b**) shows that charge varies linearly in between the electrodes but it moves to lower potentials when the relative humidity is increased (**c**) to 40% (*red curve*) and then to 80% (*blue curve*)

Still, surface properties are easily modified by intentional chemical modification, heating and other treatments including exposure to the atmosphere. Recently washed glass is highly wettable and liquid water easily spreads over it, but it becomes hydrophobic under even shorter exposure to the atmosphere. This is due to adsorption of oil that is always found in urban air or the terpenes found in the atmosphere of regions with dense vegetation.

7.2.2 *Polyethylene and Other Hydrophobic Solids*

Polyethylene (PE) surfaces undergo autoxidation, upon exposure to air. The oxidized products thus formed are hydrophilic hydroxyl, carbonyl, carboxylic and other polar groups that increase PE tendency to adsorb water. Consequently, PE should become more and more wettable with time, but this is not usually observed. The reason is the mobility of the surface groups, driven by Marangoni effect: the oxidized PE chains contribute to increase PE surface tension and they are thus spontaneously covered by non-oxidized chains migrating from sub-surface layers. This phenomenon was studied in detail by Baszkin and Saraga [3] and its knowledge is essential to understand the variability of polymer surface properties.

7.3 Charge Trapping During the Formation of Solids

Excess charge is found in the interior of some solids due to their structural features coupled to the fabrication processes. This is the case of polymer latexes, a large family of materials showing this feature that has been studied in great detail.

Persulfate ions are initiators of the emulsion polymerization process used to make latex so that the chain ends contain residual sulfate groups, contributing a negative charge. Polymer particles formed during emulsion polymerization grow up to many hundreds of nanometers and they are thus much larger than the typical radii of gyration of the individual polymer chains. The charged sulfate chain ends tend to locate at the particle–water interface but many are prevented to do so due to their size and they remain trapped within the emulsion polymer particles.

For this reason, the dry particles are spherical multipoles, as evidenced in Fig. 7.3, with a negative core whose charge derives from residual sulfate chain ends

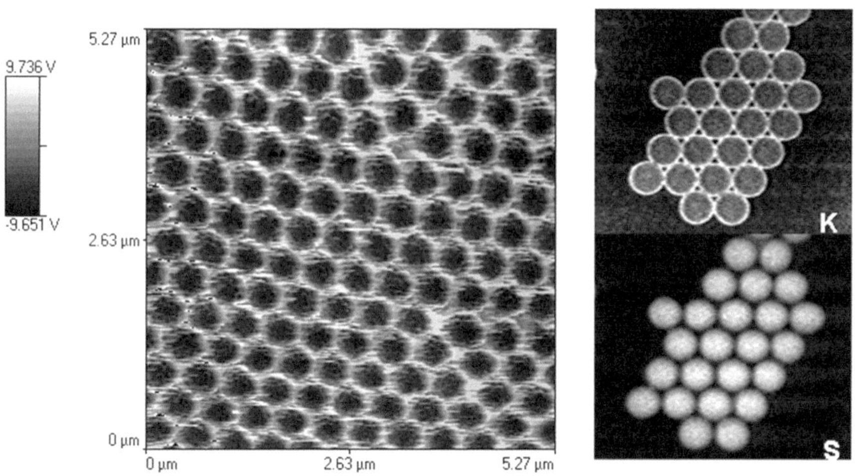

Fig. 7.3 Kelvin force micrograph (KFM, *left*) and elemental maps obtained by electron spectroscopy imaging in the transmission electron microscope (ESI-TEM, *right*) for potassium and sulfur. The sample is poly (styrene-co-hydroxyethylmethacrylate) (PS-HEMA) latex. The two kinds of images are completely independent: KFM maps electric potential while ESI-TEM provides elemental maps, but both provide the same information: the particles are core-and-shell particles, with negative cores and positive shells. Reprinted with permission from [4]

and a positive shell formed by the counter-ions [4]. The observed charge segregation is the result of a delicate balance between electrostatic interactions between the sulfate anions and potassium counter-ions, van der Waals interactions between uncharged polymer chains and hydrophobic interactions that separate water from the hydrophobic polymer chains.

An interesting consequence of negative charge trapping in the interior of latex particles is their ability to form thermoplastic solids stained with cationic dyes, like methylene blue [5]. Methylene blue does not dissolve in polystyrene, but it is incorporated within the polymer provided it is first adsorbed in the dispersed latex particles and these are later aggregated forming solid, blue polystyrene.

7.4 Electrets

The most prominent topic in this chapter is electret formation and stability, given the enormous importance of these materials in various applications and its role in the development of condensed matter physics [6]. As for the applications, electrets are reportedly present in products of the following industry sectors: materials and steel, automotive, chemical manufacturing, electronics, IT and software, telecommunications, and aerospace engineering.

Electrets are the electrostatic analogs of permanent magnets, possessing the ability to create permanent electric field in their vicinity. In practice, electrets are not absolutely permanent but the electric fields produced by them are sufficiently stable to guarantee their performance during a given time period. Electrets are different from piezo or flexoelectric materials that are electrically polarized but only under tension or compression and undergo fast relaxation, in the absence of mechanical action [7].

The possibility of the existence of electrets was already discussed by Faraday and the term "*electret*" was coined in 1885 by Oliver Heaviside [8]. In 1925 Prof. Mototaro Eguchi from the Japanese Naval College described the first permanent electret, made of an artificial membrane of beeswax, Brazilian palm gum (carnauba wax) and rosin [9]. By heating this mixture to a molten state and polarizing it within an electric field, charge was maintained for a long time afterward. This device was extensively used by the Japanese during World War II, in condenser microphones. Today, electrets are used not only in microphones but also in air filters, radiation dosimeters, electrophotography, inkjet printing and other applications [7]. A promising new application of electrets is distributed power microgeneration [10] using devices that scavenge mechanical energy from the environment converting it into electricity.

There are two main types of electrets: dipolar electrets and space-charge electrets. The dipolar electrets produce an electric field following the orientation of the molecular dipoles, thus they do not depend on excess charge. Dipolar electrets can be made from waxes or polymers.

Polarization is obtained on waxes heated above the melting point or on polymers at a temperature above the glass transition temperature (T_g) but well below the melting point. Melting is not necessary, because polymer chains are mobile above the T_g.

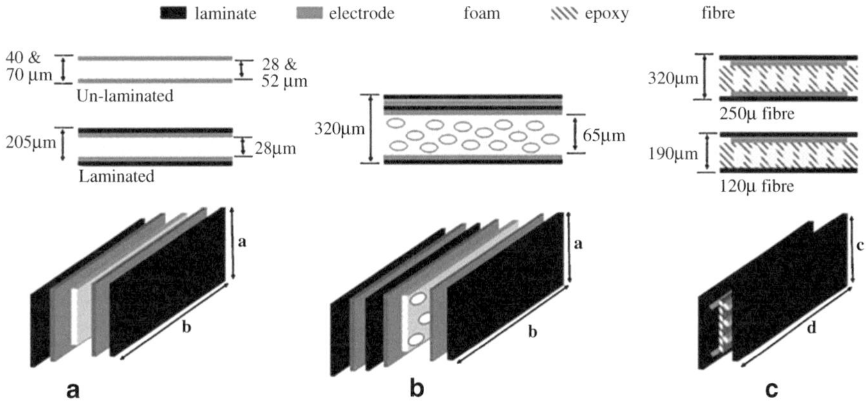

Fig. 7.4 Schematic representation of non-symmetrical piezoelectric samples (**a**) PVDF membrane film, (**b**) porous PP and (**c**) piezoelectric PZT fiber embedded in an epoxy with copper clad electrodes etched on to the inner surface of the laminate which act as electrodes. Reprinted with permission from [14]

In space-charge electrets, electric charges are added to the surface or the bulk of a material by different techniques [11] and the stability of the electret depends essentially on charge immobilization.

The performance of polymer foams as electrets is remarkable [12], due to the unique possibilities for charge separation within the foam [13]. In fact, polymers are extremely versatile materials and develop complex charge patterns, being ideal for the design of energy harvesting structures [14] such as shown in Fig. 7.4.

The charged polymer foam contains large quasi-macroscopic dipoles within the foam cells, that in turn account for piezoelectric constant as high as 250 pC/N, much higher than the widely used PVDF and close to one-half the figure for PZT. As an advantage, the foam electrets are very lightweight and flexible, as opposed to the heavy and brittle PZT [8]. Thanks to these characteristics, excitation of 15.2 cm square sample at 60 Hz and displacement of ±73 mm delivers 6.0 μW to a 1 mF storage capacitor. This reaches 4.67 V in 30 min that is suitable to supply power to low-consumption devices.

7.4.1 Thermally Stimulated Discharge

Electret stability depends largely on the immobilization of separate charge carriers or oriented dipoles that are in turn achieved by keeping the electrified solids in the crystalline or glassy states. Heating the samples allows charge motion that can be measured as current between the electret opposite faces as a function of temperature and time. For this reason, a prevalent technique to study electrets is the measurement of thermally stimulated discharge, TSD [15]. Initially developed for R&D in electrets including quality control, the technique spread to other areas like thin films, photoconductors, electro-optical devices and ice microcrystals [16].

7.4.2 Bioelectrets

Many biomaterials display the ability for charge and polarization storage. Moreover, biological functions are always performed in the presence of water that is often referred to as holding special properties, in biomaterials and contributes to materials charging, in many different ways as discussed in Chap. 6.

Electret applications in biological systems include anti-thrombogenic surfaces, the stimulation of tissue growth in bone and special artificial membranes [17].

7.5 Charging Mechanisms

7.5.1 Unintended Charging of Solids, Following Mechanical Action and Radiation

There are many situations where charges can be formed in bulk solids following mechanical action, either by implantation or trapping as part of the relaxation mechanisms. Free energy of mechanically stressed solids is higher than in the relaxed state and excess mechanical or elastic energy may produce structural faults, especially of the dislocation type [18]. When sufficiently large gradients of the chemical potential are created, the excess energy may trigger the formation of a triboplasma [19], a collection of high-energy chemical species. These further evolve through a host of parallel and consecutive reactions taking place in the surface but also in the interior of the solid, analogous to reactions triggered by high-energy particles and radiation [20].

Sample charging in electron microscopes and photoelectron spectrometers is familiar to practitioners of these techniques and special measures are taken to avoid undesirable effects. For instance, samples for scanning electron microscopy are often coated with carbon or metal thin films. Transmission electron microscope samples are supported on metal grids and isolated particles are held on thin carbon films, to facilitate the dissipation of the electron current. Photoelectron spectrometers are usually fitted with ion flooding devices, to avoid charge build-up on dielectric samples that introduces errors in the energy scale of the acquired spectra.

Mechanochemical and radiation or beam-induced reactions produce charged species. As a result, "charge storage effects are observed in most instances when solid dielectrics are exposed to penetrating radiation", following B. Gross [21]. For instance, the formation of free radicals during the rupture of chemical bonds is followed by electron-transfer reactions that lead to the formation of charged species [22]. The result may be one or more of various physical effects: micro-electrostatic discharges, triboluminescence, phonon and heat propagation [23]. This topic is further developed in Chap. 9 in this book.

Since every material on Earth surface is constantly bombarded by radiation stemming from the outer space, from the atmosphere of minerals and/or is subjected to various kinds of mechanical action, any material is constantly producing charge species in its interior. If these cannot migrate fast, the interior of the solid will store charge.

Many other phenomena are used to implant charges in solids, purposely. These are shortly described ahead and detailed information on this subject can be found in the book edited by Sessler and Gerhard-Multhaupt [24] and related publications.

7.5.2 Charging with Corona

A very successful way to inject charge carriers is *corona charging*. Corona is ionized gas produced under an inhomogeneous electric field sufficient to provoke the formation of ionized, high-energy species in the gas phase but still below the discharge potential. The formation of unstable species is evidenced by the emission of (usually) bluish light around the electrode with the smaller area that is the visible "corona". Usually, corona charging is achieved by polarizing a needle right above the sample surface laid over a grounded flat support. Increasing the potential difference also increases the rate of corona charging. If the needle is negatively biased, negative carriers are formed and they flow toward the dielectric while still undergoing further reactions. In air corona, negative species are primarily CO_3^- [24]. Beyond the species formed in the atmosphere further reactions may take place within the charged polymer, as evidenced by the improvements in electret performance when some additives are added to polypropylene and polycarbonate films and also when the films age [25, 26]. The simplicity of corona charging makes it appropriate to use in large-scale fabrication. It has been extensively used in the production of electrets for microphones and is also widely used in xerography.

7.5.3 Direct Charge Injection from Electrodes

Charge injection is widely used in electrospinning and electrospray techniques. As schematically shown in Fig. 7.5, when a solution or some pure liquid is passing through a polarized metal needle (in the range of kilovolts), the electric field between the needle and a grounded collected plate is on the order of few thousand kV per centimeter [27]. Electrospinning produces a jet of electrified liquid while the electrospray is a jet of charged droplets, depending on the internal pressure of the liquid and other factors like viscosity. Electrospinning method is usually applied to polymer solutions or polymer melts resulting in nanoscaled fibers and electrospray is a conventional technique used in mass spectrometry. Moreover, the charging mechanisms here are the same as described for water in Chap. 6.

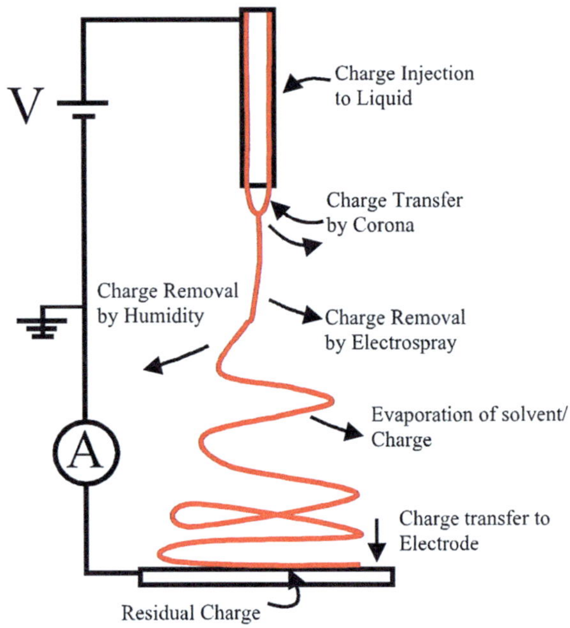

Fig. 7.5 Schematic diagram of a typical electrospinning experiment. Reprinted with permission from [27]

7.5.4 Electron, Ion and Photon Beams

The implantation of negative charges can be also done by impinging electron or ions beams on the sample. The incident particles contribute their own charges and also the charges formed by any one of the various inelastic collisions that they can undergo. Particles travelling through the bulk solid are thus slowed down becoming finally trapped and contributing their charges. Similarly, *penetrating radiation* like X-rays provokes many effects including electron ejection, leaving behind cations and consequently forming positive electrets.

Drum charge control in xerography is done by using visible light that triggers photoconduction, discharging the desired pixels and thus creating a latent image.

There is excellent literature on this topic that will not be further detailed here.

7.5.5 Electrification by the Liquid Contact Method

Charge transfer between a liquid and an insulating solid under an applied electric field was used in electrostatic recording and electret production with different liquids and insulating materials. Charge transfer measurements and a detailed analysis of this method were presented by Chudleigh [28]. Water, ethanol and dilute solutions of HCl and NaOH were placed in between an electrode and the dielectric surface of a fluorinated polymer film, while the other film side is metal-coated and connected to ground. A schematic description is given in Fig. 7.6.

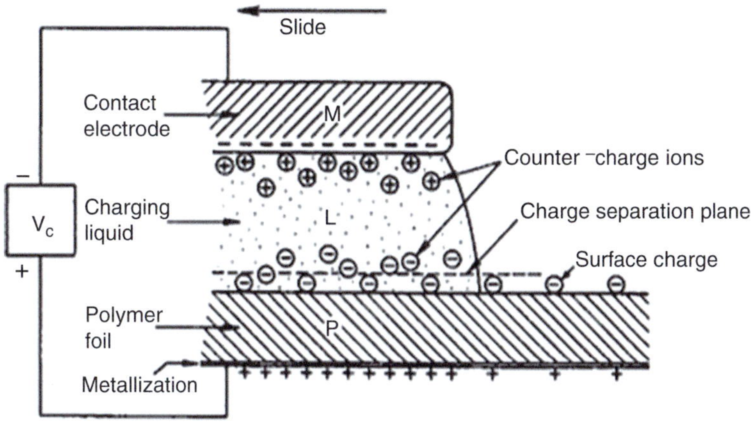

Fig. 7.6 Schematic diagram of apparatus for charging a polymer film using a liquid contact. Applying a negative charging voltage produces a negative surface charge on the polymer film. Reprinted with permission from [28]

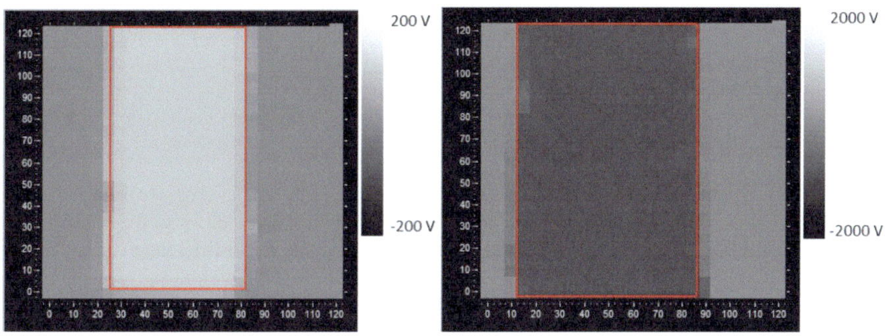

Fig. 7.7 Electrostatic potential maps of PP/aluminum backside-coated films charged using Chudleigh's method. Electrode was polarized with (*left*) +100 V and (*right*) −1000 V

When a potential difference is applied between the electrode and the metal-coated surface, charge double-layers form at both solid–liquid interfaces and they are broken when the liquid is withdrawn, leaving behind excess charge on the polymer film.

For example, a polypropylene (PP) film coated on one side with aluminum was charged using Chudleigh's method, by wetting its surface with cotton soaked on ethanol biased to the desired voltage, while the aluminum-coated surface was grounded [29]. The PP area was 8 × 12 cm^2 divided into 5 × 5 mm^2 spots that were contacted by the biased wet cotton for 5 s each, while this was displaced by a mechanical arm. After the completion of the scan, the electrode was removed while still biased. Then, the region was scanned with a Kelvin electrode kept at 2 mm above the surface for surface potential mapping as shown in Fig. 7.7. The map shows that charge acquired using the liquid-contact method is highly uniform, since the surface potential is uniform within ±1 V.

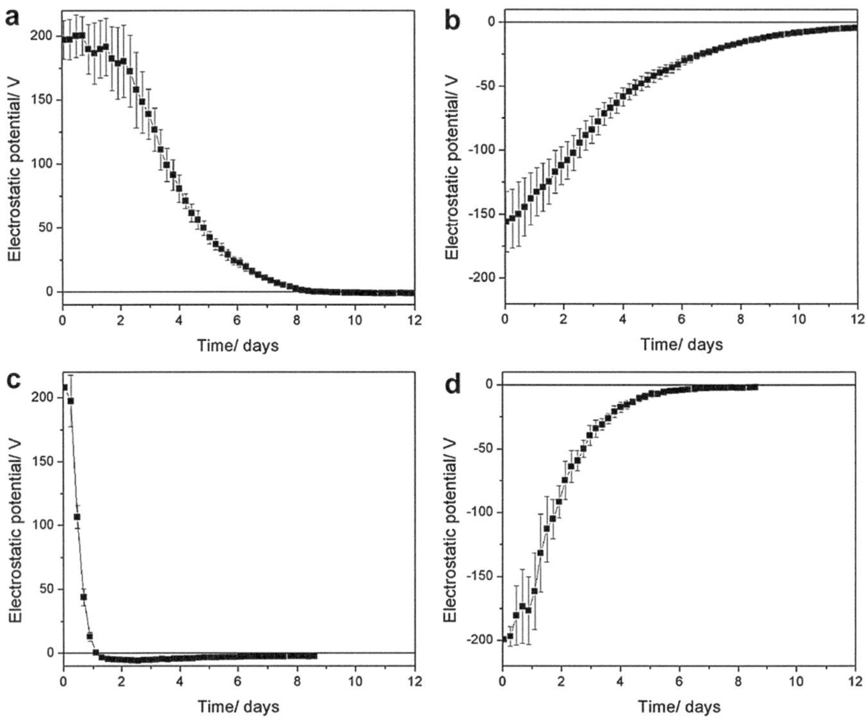

Fig. 7.8 Electric potential decay as a function of time on polyethylene samples: (**a**) and (**b**) at 3% RH; (**c**) and (**d**) at 60% RH. Averaged values with error bars are obtained from the ten different pixels on each PE sample. Note that (a) and (c) are initially positive and (b) and (d) are negative. Reprinted with permission from [30]

7.6 Charge Migration from Charged Solids

Excess charge in solids tends to migrate to surrounding environment and to the ground. The mechanism that has been almost exclusively considered for charge migration is electric conduction through the surrounding bodies but recent results and mechanistic proposals produced a richer picture.

Polyethylene (PE) exposed to corona discharge acquires positive or negative charge that may be scanned with a Kelvin probe. The resulting data are presented as an electrostatic potential map and successive scans show changes in the map, giving detailed information on charge patterns as well as on the kinetics and destination of charge migration. Figure 7.8 shows potential maps of corona-charged PE surfaces and potential vs. time plots for chosen pixels of the maps [30]. These results revealed some surprising information that is discussed in the following paragraphs.

Corona-charged polyethylene samples follow exponential decay and the dissipation rates are usually slower at a 3% relative humidity than at higher RH values (20, 40 and 60%) but the values also vary with the potential sign. Negative potentials

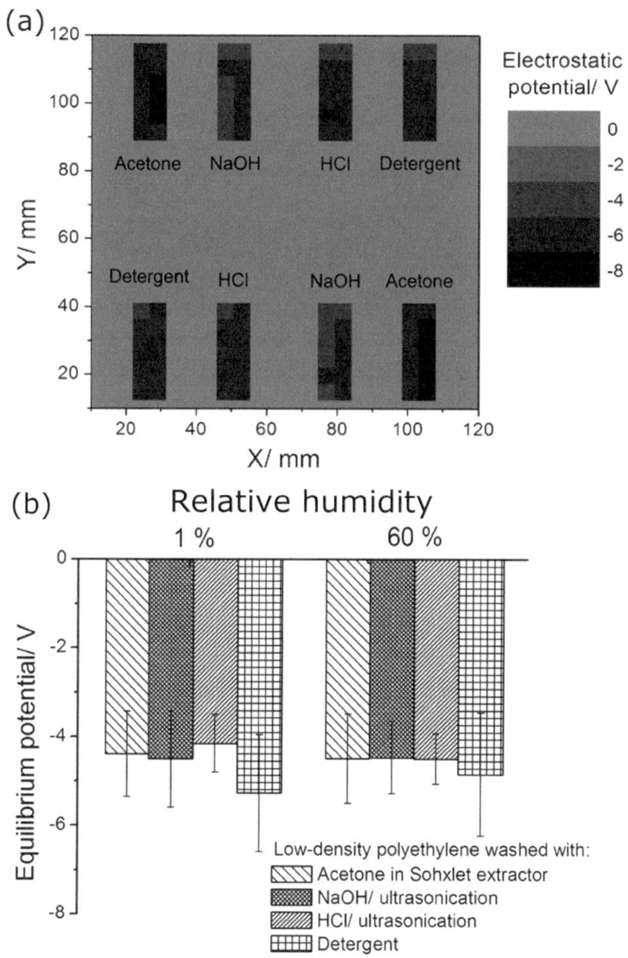

Fig. 7.9 (**a**) Equilibrium potential maps of polyethylene samples washed in different ways at 60% RH; (**b**) averaged equilibrium potentials at 1 and 60% RH. Reprinted with permission from [30].

decay at a slower rate than positive ones, which is especially noticeable under relative humidities higher than 20% [30]. Also, there is a difference between positive and negative charge decay: for positive potentials, half-lives are a few times longer under 3% RH than under 60% RH, while the average half-life for dissipation of negative potentials is only ca. two times longer at 3% RH than at 60% RH.

Moreover, when PE samples are initially charged with corona, either positive or negative, the potential varies with time reaching small negative values, -4.6 ± 0.7 V, as shown in Fig. 7.9. Thus, there is an *equilibrium potential* for polyethylene, whose generality has not yet been verified in other cases. One possible explanation for this non-zero potential is preferential adsorption of OH^- ions from atmospheric water. Another possibility is H^+ ion desorption from carboxylic acid groups formed by oxidation of the polymer surface. Whatever explanation will prevail, this is conceptually very important because it shows that non-zero potential is the equilibrium state, as opposed to the common assumption of electroneutrality of pure substances.

Ieda and collaborators [31] found that corona-charged polyethylene samples with high potential decay more rapidly than other samples charged to a lower potential. This may be explained by considering the existence of deep charge traps on the surface and shallow traps in the bulk: high-energy charged species formed under high voltage can move easier into the bulk solid, occupying shallow traps. However, charged species formed at low corona voltage are unable to migrate to the bulk or even along the surface and are thus caught in deep surface traps [32]. The nature and chemical or structural features of deep and shallow traps are as yet unknown [33, 34] and it is likely that their identification will be achieved but only by using various sophisticated analytical tools, analogous to the case of the identification of charged species formed by friction [22].

7.7 The Costa Ribeiro (Thermo-Dielectric) Effect

Joaquim Costa Ribeiro was a pioneer in Brazilian physics, interested in the study of electrets. During his experiments, he discovered that an electret may be formed in the absence of an external electric field, during melting or crystallization [35] of carnauba wax, the same material that was used by Eguchi. This unexpected result showed that charge partition takes place during phase transitions. Ribeiro's first publication in 1943 was written in Portuguese and it was virtually ignored outside Brazil. It was later published in English, at the same time of Workman and Reynolds publication [36] on electret formation during ice solidification. In this case, potential differences as high as 230 V were observed during freezing.

This effect was interpreted by Gross [37], as follows: when the interface is displaced during solidification or crystallization, a charge gradient is created producing a potential different across the interface. The excess charges are then trapped in the solidified material, transforming it in an electret.

The thermo-dielectric effect was also related by Reynolds to charge separation in thunderstorms [38]. It is especially important in the case of dendrite formation from super-cooled water [39] and it may be seen as part of the broader topic of charge separation during phase transition [40].

7.8 Conclusion

Solids acquire excess charge following many different mechanisms that are largely dependent on the chemical and structural characteristics of the solids involved. Charge is found in the interior of the solids as well as at their surfaces, thus the overall charge depends on the surface properties of the solid, that are highly variable, in many important cases as e.g. in thermoplastics. The abatement of the charge measured on a solid and of the electric potential measured in its vicinity may reflect the migration of charge carriers away from the solid but perhaps more often it is the consequence of the adsorption of ions from the atmospheric water.

References

1. Barthel AJ, Kim SH (2013) Surface chemistry dependence of water adsorption on solid substrates in humid ambient and humidity effects on wear of copper and glass surfaces. Tribol Mater Surf Interf 7:63–68
2. Gouveia RF, Costa CAR et al (2008) Water vapor adsorption effect on silica surface electrostatic patterning. J Phys Chem C 112:17193–17199
3. Baszkin A, Ter-Minassian-Saraga L (1971) Chemical structures of surface-oxidized and grafted polyethylene: adsorption and wetting studies. J Polym Sci Pt C 34:243–252
4. Rezende C, Gouveia RF et al (2009) Detection of charge distributions in insulator surfaces. J Phys Condens Matter 21:263002
5. Braga M, Leite CAP et al (2003) Hydrophobic polymer modification with ionic reagents: polystyrene staining with water-soluble dyes. Langmuir 19:7580–7586
6. Sessler GM (1987) Physical principles of electrets. In: Sessler GM (ed) Topics in applied physics, vol 33. Springer, Berlin, pp 13–80
7. Kaufman GK, Gooding DM (2014) Electrets. In: Encyclopedia of inorganic and bioinorganic chemistry. Wiley, New York, pp 1–10
8. Heaviside O (1885) Electromagnetic induction and its propagation. Electrician 14:230
9. Eguchi M (1925) On the permanent electret. Philos Mag Ser 49:178–192
10. Zhou T, Zhang L et al (2016) Multilayered electret films based triboelectric nanogenerator. Nano Res 9(5):1442–1451
11. McCarty LS, Whitesides GM (2008) Electrostatic charging due to separation of ions at interfaces: contact electrification of ionic electrets. Angew Chem Int Ed 47:2188–2207
12. Gerhard-Multhaupt R (2002) Less can be more. Holes in polymers lead to a new paradigm of piezoelectric materials for electret transducers. IEEE Trans Dielectr Electr Insul 9:850–859
13. Anton SR, Farinholt KM et al (2014) Piezoelectric foam-based vibration energy harvesting. J Intell Mater Syst Struct 25:1681–1692
14. Patel I, Siores E, Shah T (2010) Utilisation of smart polymers and ceramic based piezoelectric materials for scavenging wasted energy. Sens Actuat A Phys 159:213–218
15. van Turnhout J (1987) Thermally stimulated discharge of electrets. In: Sessler GM (ed) Topics in applied physics, vol 33. Springer, Berlin, pp 81–215
16. Pissis P, Apekis L et al (1987) Multiplicity of dielectric relaxation times of dispersed ice microcrystals. Time dependence. IL Nuovo Cimento D 9(2):195–211
17. Mascarenhas S (1987) Bioelectrets: electrets in biomaterials and bioplolymers. In: Sessler GM (ed) Topics in applied physics, vol 33. Springer, Berlin, pp 321–346
18. Vasiliu-Oprea C, Dan F (2006) Macromolecular mechanochemistry: polymer mechanochemistry. Cambridge International Science, Cambridge, p 390
19. Heinicke G (1984) Tribochemistry. Carl Hanser Verlag, München—Wien
20. Wojnárovits L (2011) Radiation chemistry. In: Vértes A, Nagy S et al (eds) Handbook of nuclear chemistry. Springer, Berlin, pp 1263–1331
21. Gross B (1987) Radiation-induced charge storage and polarization effects, Chapter 4. In: Sessler GM (ed) Electrets, topics in applied physics, vol 33. Springer, Berlin
22. Burgo TAL, Ducati TRD et al (2012) Triboelectricity: macroscopic charge patterns formed by self-arraying ions on polymer surfaces. Langmuir 28:7407–7416

23. Burgo TAL, Erdemir A (2014) Bipolar tribocharging signal during friction force fluctuations at metal-insulator interfaces. Angew Chem Int Ed 53:12101–12105
24. Sessler GM (1987) Physical principles of electrets. In: Sessler GM (ed) Topics in applied physics, vol 33. Springer, Heidelberg. Chapter 2, pp 13–80
25. Mohmeyer N, Müller B et al (2004) Nucleation of isotactic polypropylene by triphenylamine-based trisamide derivatives and their influence on charge-storage properties. Polymer 45:6655–6663
26. Erhard DP, Lovera D et al (2010) Tailored additives to improve the electret performance of polycarbonates. Macromol Chem Phys 211:2179–2186
27. Collins G, Federici J et al (2012) Charge generation, charge transport, and residual charge in the electrospinning of polymers: a review of issues and complications. J Appl Phys 111(4):044701
28. Chudleigh PW (1976) Mechanism of charge transfer to a polymer surface by a conducting liquid contact. J Appl Phys 47:4475
29. Burgo TAL, Silva CA et al (2013) Friction coefficient dependence on electrostatic tribocharging. Sci Rep 3:2384
30. Burgo TAL, Rezende CA et al (2011) Electric potential decay on polyethylene: role of atmospheric water on electric charge build-up and dissipation. J Electrostat 69:401–409
31. Ieda M, Sawa G et al (1967) A decay process of surface electric charges across poyethylene film. Jpn J Appl Phys 6:793–794
32. Xu Z, Zhang L et al (2007) Decay of electric charge on corona charged polyethylene. J Phys D Appl Phys 40:7085–7089
33. Wintle HJ (1972) Surface-charge decay in insulators with non-constant mobility and with deep trapping. J Appl Phys 43:2927–2930
34. Baum EA, Lewis TJ et al (1977) Decay of electrical charge on polyethylene film. J Phys D Appl Phys 10:487–497
35. Ribeiro JC (1950) On the thermo-dielectric effect. Anais Acad Bras Cienc 22:325–347
36. Workman EJ, Reynolds SE (1950) Electrical phenomena occurring during the freezing of dilute aqueous solutions and their possible relationship to thunderstorm electricity. Phys Rev 78(3):254–259
37. Gross B (1954) Theory of thermodielectric effect. Phys Rev 94(6):1545–1551
38. Reynolds SE, Brook M et al (1957) Thunderstorm charge separation. J Meteorol 14:426–436
39. Bauerecker S, Buttersack T (2014) Electric effect during the fast dentritic freezing of supercooled water droplets. J Phys Chem B 118:13629–13635
40. Amin MS, Peterson TF et al (2006) Measurements of electric charge associated with evaporation and condensation of water on metallic surfaces as a consequence of pressure, humidity, and temperature change. J Electrostat 64:597–603

Chapter 8
Friction and Electrostatics

Contents

8.1 Introduction

Friction and electricity are closely related and the latter is a common outcome of the former, as much as wear. Attractive forces between rubbed solids have been noticed since ancient times and many scientists claim that *triboelectrification* (electrification by friction) is the oldest known manifestation of the electrical sciences. On the other hand, although both tribology and electrostatics are related to everyday phenomena, many basic issues are still matters of debate and did not receive much attention during most of twentieth century. However, a surge of progress has been observed in both cases for the past few decades and the mutual feedback between friction and charging is now clear.

Tribology and electrostatics are inherently complex sciences, thus it is not surprising that the their mutual relationships are also complex. The conception that

F. Galembeck, T.A.L. Burgo, *Chemical Electrostatics*,
DOI 10.1007/978-3-319-52374-3_8

friction must be related to attractive forces seems to be very intuitive: whenever two bodies are mutually attracted, it must be difficult to slide one of them over the other. In 1734, Desaguliers observed that when metals were polished, this could increase friction due to adhesive forces. However, his observations contradicted the *Amontons' law* that was then already established. This law states that friction coefficients are independent of the contact area or, in other words, friction forces are only dependent on the normal load and proportional to it. Although Desaguliers' experimental findings were correct, he could not explain why friction coefficient is independent of the contact area while adhesion is dependent.

Much later in 1950, Bowden and Tabor clarified this apparent contradiction by introducing the concept of *real contact area* that is the summation of the areas of numerous microscopic regions where asperities from both contacting surfaces form contacting junctions. These contacting sites may be more or less numerous, depending on surface roughness and they are the only sites for atom-to-atom contact [1]. These authors showed that the frictional force is strongly dependent on the real contact area and increasing the normal force also increases the number and extent of contacting sites. They could thus explain why adhesive interactions contribute to frictional forces, although real area of contact was then more a qualitative notion than a measurable property of the system.

8.1.1 Adhesive Contact Models: JKR, DMT and Maugis

In his model for mechanical contact, Hertz [2] assumed that only compressive stresses at the interface were responsible for the contact between solid elastic bodies and he found that the area of contact varies non-linearly with the normal force since it is proportional to $F^{2/3}$. At the contact zone, a pressure distribution is created that elastically deforms the interface. Moreover, this pressure distribution $p_{(r)}$ is given by Eq. 8.1. It is semielliptical over the contact area peaking at its center:

$$p_{(r)} = p_0\left(1 - r^2 / a^2\right)^{\frac{1}{2}} \tag{8.1}$$

where a is the radius of the contact zone, r is the distance from the center of the contact to the point being considered ($r^2 = x^2 + y^2$) and p_0 is the maximum pressure. The Hertz analysis is fairly universal and routinely used in the analysis of tribology experiments. On the other hand, interactions between solids, particularly between colloidal particles, were later established as the result of interatomic forces, especially of van der Waals forces. It was then found [3] that an extra force is needed to separate two solids in close contact. Although Bowden and Tabor could experimentally verify that adhesion contributes to friction due to elasto-plastic deformations at the interface, it was only during the 70s that a great effort was made to understand the contribution made by adhesive forces to Hertzian contact.

In a very elegant work, Johnson, Kendall and Roberts [4] (JKR were capable to include elastic deformations arising from the adhesive forces to the contact equilibrium between elastic bodies. They assumed that adhesive forces (at the interface only) increase the contact area and change both penetration and stored elastic energy. So, if a tensile stress induced by pulling forces is added to Hertzian compressive stress, the total normal stress $p(r)$ at the interface is:

$$p(r) = p_0\left(1 - r^2 / a^2\right)^{\frac{1}{2}} + p_0'\left(1 - r^2 / a^2\right)^{-\frac{1}{2}} \tag{8.2}$$

In the Hertzian theory, a negative value of p_0' was rejected because tension could not be sustained under negative loads. The first term in Eq. (8.2) is positive, while the second, the tensile stress induced by surface forces, is negative. So, in the presence of adhesive forces the tensile stress deforms the edges of the interface and consequently increases the contact area. JKR theory is especially successful with elastic bodies under low loads. On the other hand, it fails under high loads, when the experimental results are well fitted by Hertz theory. This observation suggests that short-range surface forces only become important when the normal load tends to zero, for the contact of elastic materials. The surface stress distribution predicted by Eq. (8.2) is such that stress goes to infinity at the edge of the interface. On the other hand, this very high stress is predicted but not observed, because JKR theory implicitly assumes that surface forces act over an infinitesimally small distance exactly at the interface and not outside the contact area [5].

Both theory and successful verification of the original JKR theory are based on short-range attractive forces and it is thus especially valid for smooth surfaces with low elastic modulus because these short-range surface forces are not significant at distances exceeding some nanometers. Moreover, soft materials (in the original work the authors used rubber over gelatin) can be flattened to fill in asperities, forming a very intimate interface. On the other hand, for more realistic conditions as when the interface is not much deformed and materials are sufficiently rigid, the stress profile can be assumed as Hertzian.

In fact, Derjaguin, Muller and Toporov [6] (DMT) realized that for rigid materials, adhesive forces cannot overcome elastic modulus and consequently the interface is not the result of deformed surfaces. Then, adhesive forces are taken into account but they must be computed outside the contact area so that the normal load is the result of the external load and the adhesion force. In practice, the adhesive force measured with an indenter is:

$$F_{(\text{adhesion})} = 2\pi R W \tag{8.3}$$

where R is the spherical indenter radius (for a sphere-on-flat geometry) and W is the work of adhesion. $F_{(\text{adhesion})}$ is then added to the Hertz contact equation (Eq. (8.1)). Many practical situations fit well within DMT theory. For instance, the Quantitative Nanomechanical mode (QNM®) of AFM, recently developed by Bruker, uses the

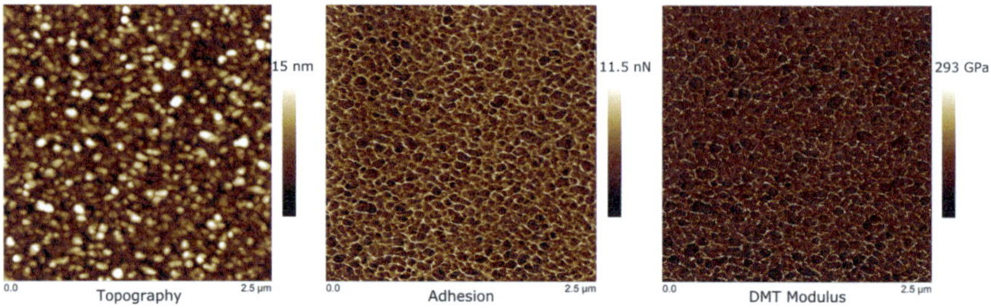

Fig. 8.1 Application of DMT theory. Quantitative nanomechanical mode (QNM) for plasma-deposited nickel nitride (Ni-N) film. The Young modulus is calculated from *DTM analysis*

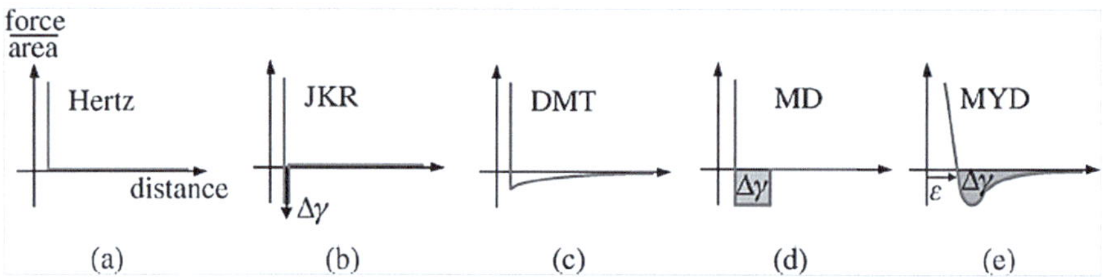

Fig. 8.2 Description of contact mechanics models. While the Hertz model does not consider the adhesion in contact, the JKR model includes only a short-range adhesion in the contact area and is a function of the work of adhesion $\Delta\gamma$. The DMT model shows a long-range surface forces and the Maugis-Dugdale (MD) model considers the square well potential to describe attractive forces. The Muller, Yushenko and Derjaguin (MYD) uses a Lennard-Jones potential to include both the short-range and long-range forces. Reprinted with permission from [10]

DTM theory to calculate the Young modulus of materials during force-curve measurements. Figure 8.1 shows the topography, adhesion and DMT modulus mappings for a plasma deposited nickel-nitrate (Ni-N) film. The adhesion mapping is obtained from pull-off force. On the other hand, the Young modulus is obtained by fitting the retract curve to the DMT model. Note that in Fig. 8.1, most regions where adhesion is high also show a high DMT modulus, as well.

Although controversy and heated debates on the applicability of both theories were entertained for some years, Tabor [7] realized that the two theories represent the opposite extreme cases of mechanical contact: JKR is more applicable to large radius, compliant solids with strong adhesion while DMT is more indicated for small radius, rigid solids with low adhesion. Later, using a Dugdale (square well) potential model to describe attractive forces, Maugis [8] derived analytical solutions to show the transition between both theories and he was capable to handle intermediate cases. Figure 8.2 shows a comparison between some important models. In practice, the Hertz analysis and JKR-DMT theories are the most used models because they use simple equations relating the normal stress with the radius of the contact zone. For more precise work, the Maugis-Dugdale (MD) or even the Muller, Yushchenko and Derjaguin (MYD) model [9] (which describes the adhesion force via Lennard-Jones potential) can be used [10].

There are many articles and books where all derivations of the above-mentioned models and equations can be found in detail. In this chapter, we only describe briefly the main features of the most widely used theories to contextualize tribocharging within the main problems in contact mechanics.

Study in this area produced outstanding results of the direct measurement of van der Waals forces acting between surfaces, thanks to the famous Surface Forces Apparatus (SFA) developed by Tabor, Winterton and Israelachvili in the early 1970s.

At this point, it is worth stressing that *friction* does not correlate directly with *adhesion*. This seems contradictory, but there are many examples of bodies with high friction coefficients but having low adhesive forces, and vice versa. On the other hand, since friction is an energy-dissipation process it actually relates well to *adhesion hysteresis* [11], a concept introduced by Israelachvili [12]: the work required to separate two surfaces is typically greater than the work done when contacting them. Adhesion hysteresis is the difference between both and friction can be predicted from the adhesion hysteresis curve.

8.2 From Macro to Nanoscale

For macroscopic objects, forces such as gravity tend to be more significant than those forces (van der Waals or Coulombic) that arise from intermolecular interactions. Also, friction has many factors other than surface forces, like cohesion, roughness, viscoelasticity, hardness, lubricated conditions, elasto-plastic deformability and perhaps still others.

The basic laws of friction established by Amontons and Coulomb are three empiric laws:

first law: friction is proportional and perpendicular to the normal load;
second law: friction is independent of the apparent contact area;
third law: kinetic friction is independent of sliding velocity.

Although these laws may appear simplistic and they are not derived from fundamental principles, they are extremely valuable for solving real life problems. Moreover, the relationships stated by these laws are the result of many important interfacial, molecular and materials phenomena that are by themselves rather complex or not fully understood.

Amontons already noticed that uneven surfaces have higher friction coefficients than smooth surfaces. Roughness is thus an important variable that lead many authors to think that friction was originated only from roughness and nothing else.

On the other hand, after Bowden and Tabor introduced the concept of real contact area, the comprehension of friction phenomena reached a higher level. Today, it is accepted that friction has two basic factors [5]:

1. The *adhesion force* needed to overcome attractive interactions developed through the outermost parts (tips of the asperities) of the surfaces;
2. The *plowing force* that is observed when one surface is significantly harder than the other. In this case the softer surface is plowed by the asperities on the harder one [13].

Friction force can be written as:

$$F = F_{\text{adhesion}} + F_{\text{plowing}} \tag{8.4}$$

Both adhesion and plowing force are by themselves the result of many important surface phenomena. Attractive interactions (van der Waals forces), mechanical shearing, elasto-plastic deformations, contaminants, water adsorption, cohesion (weakly bound layers [14]), and presence of third bodies [15] (material transfer) are only some of many factors that can influence the adhesion and plowing forces. On the other hand, at the nano (or micro) scale surface forces can overcome many of the energy dissipating mechanisms so as to become more important than these as well as gravity.

Progress in the relevant instrumentation, SFA and AFM allowed the demonstration of the importance of coulombic interactions between surfaces to friction across interfaces under relative motion, at the microscopic level.

Salmeron and collaborators applied bias voltage to a *n*-type Si(100) wafer with stripes of highly B-doped p-type semiconductor, to control friction forces measured while passing the AFM tip over the wafer surface. They could thus change friction as the result of interactions between excess charge in the wafer and the tip [16, 17].

Reduction of friction forces on high-temperature superconductors was achieved due to the underlying atomic-scale electronic and phononic mechanisms [18, 19]. Instead, when the resulting forces across the interface are attractive, friction is raised and persists even under negative loads [20].

It is expected that forces from asymmetric fluctuating electric fields that are formed as the result of triboelectrification of moving bodies dissipate energy by dragging charge along the surface and thus increasing friction losses [21]. This statement has not yet been experimentally proven.

8.3 Electrostatic Contribution to Friction

Coulombic interactions are long-range forces that extend farther than other surface forces. Usually, van der Waals forces do not extend beyond few tens of nanometers falling as d^{-7} with the separation distance, d. On the other hand, electrostatic forces are felt even at macroscopic distances and they should not continue to be neglected in contact mechanics and friction. Considering that the electrostatic field developed at dielectric interfaces can be as high as to produce air dielectric breakdown, static electricity effects must be included in comprehensive atomic-scale treatment of friction, especially when insulating materials are involved.

The only explanation for the current neglect of Coulombic interactions in friction is the enormous complexity introduced by considering electrostatic forces. First, there is still a widespread lack of understanding of the mechanisms and patterns of electrostatic charging of materials, at the atomic-molecular level, that is discussed in other chapters in this book. Also, electrostatic interactions at the solid–gas interface are strongly dependent on many ambient factors, including its water content [22] that in turn depends on the relative humidity. Moreover, even approximate amounts of charge carriers formed or exchanged during contact are hardly known, when one of the contacting bodies is an insulator [23].

Nakayama and collaborators [24] have shown that the extremely intense electric fields generated due to tribo-physical and –chemical phenomena produced by sliding contacts create sites with high energy density, where a *triboplasma* is formed. A spectacular demonstration of the formation of high-energy species was made by the Putterman group, showing the emission of Bremsstrahlung X-rays [25] produced under intense electric fields that were in turn the result of surface charging by breaking adhesive bonds.

Budakian and Putterman investigated the effect of charge transfer in stick-slip friction at metal–insulator interfaces [26]. They found that charge transfer accompanying the slip events is proportional to the force jumps, pointing toward a common origin for triboelectrification and friction, at least in their system. The effect of triboelectricity on friction coefficients of metals was also demonstrated in different systems [27–29] but not in insulator–insulator interfaces, where quantitative information on the amount of charge and its distribution is hardly available.

8.3.1 Macro Experiments Relating Surface Charge and Friction Coefficients

Surprisingly, the effect of triboelectricity on friction coefficients was only recently quantified, for two dielectric solids at the macroscale and under well-defined conditions. This was only possible when the author's group had already accumulated 15 years of experience in the investigation of electrostatic charging of materials, including techniques for potential and charge mapping from the nano- to the macroscale [22].

This allowed the production and assessment of insulator surfaces with fairly uniform charge density [22, 30]. Sliding felt wool over a PTFE sheet produces negatively charged surfaces with lower local variations than most other systems. Half-life for potential decay in these surfaces is in the range of a few days so that samples can

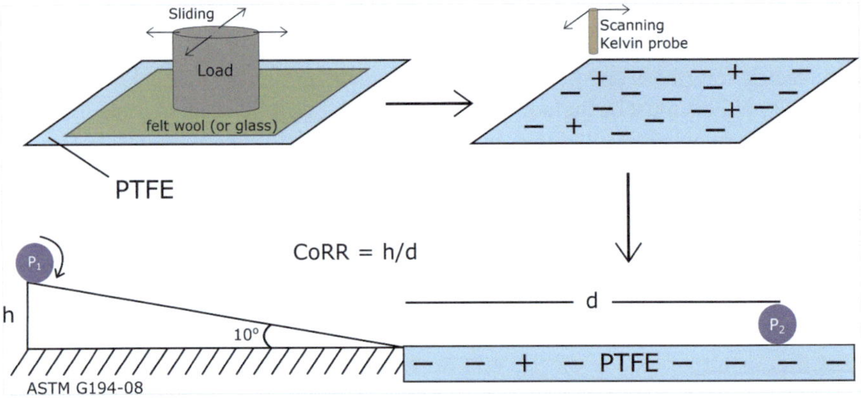

Fig. 8.3 Coefficient of rolling resistance on tribocharged surface setup. PTFE was supported on an aluminum holder and rubbed with felt wool (14 × 14 cm^2) with pressure adjusted to 1.0 ± 15 kPa that was slid for 4 cm at the speed of 1 cm s^{-1}. CoRR measurements followed the ASTM Standard (G194-08)

be prepared, and surface static potential distribution can be mapped. Most uniform samples covering a range of charge densities are thus selected for friction measurements, as shown in Fig. 8.3.

The experiments to determine the coefficient of rolling resistance (CoRR) of glass beads on PTFE as a function of the electrostatic potential in the latter (previously mapped with a Kelvin probe) are straightforward: glass beads are placed at a chosen height (in this case, $h = 1.25$ mm) on an aluminum smooth ramp with a suitable inclination angle (10°) and then allowed to roll down onto a flat tribocharged PTFE surface. The distance roamed by each sphere on the PTFE surface, d, is used to calculate the coefficient of rolling resistance, CoRR = h/d.

Figure 8.4 shows that rolling friction of glass spheres increases with electrostatic potential on charged PTFE surfaces [31]. The electrostatic potential goes from values close to 0 V to −3300 V in the various samples (Fig. 8.4a). Neutral glass spheres released from height h roll over charged PTFE but they quickly stop, after moving for only a fraction of the distance observed in uncharged PTFE. This shows that charge on the films introduces a powerful mechanism for mechanical energy dissipation. These experiments allow the calculation of the coefficient of rolling resistance (CoRR), as shown in Fig. 8.1b, as a function of the average potential on the film. CoRR increases many-fold in charged PTFE, in a potential range that is easily achieved by rubbing this polymer with glass and other common materials.

The effect of tribocharging on sliding friction was examined by sliding polyethylene (PE) pellets on PTFE film, when they are immobilized by triboelectrification. The experimental setup can be seen in Fig. 8.5: PE pellets were randomly placed on PTFE surface and shaken on a reciprocating table for 5 min. Then, the holder

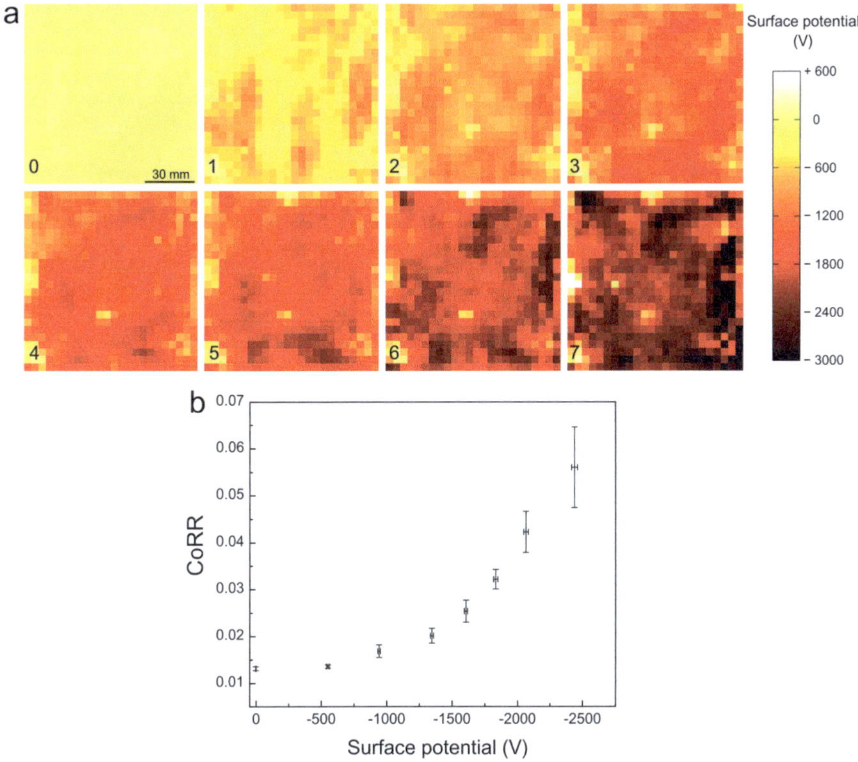

Fig. 8.4 Determination of CoRR of glass beads rolling on tribocharged PTFE surfaces. (**a**) The potential map for each plate used. (**b**) CoRR vs. average surface potential of tribocharged PTFE plates. Vertical error bars are mean standard deviations from ten replicate measurements while the horizontal bars are standard deviations of average potential for all the pixels on each plate. Reprinted with permission from [31]

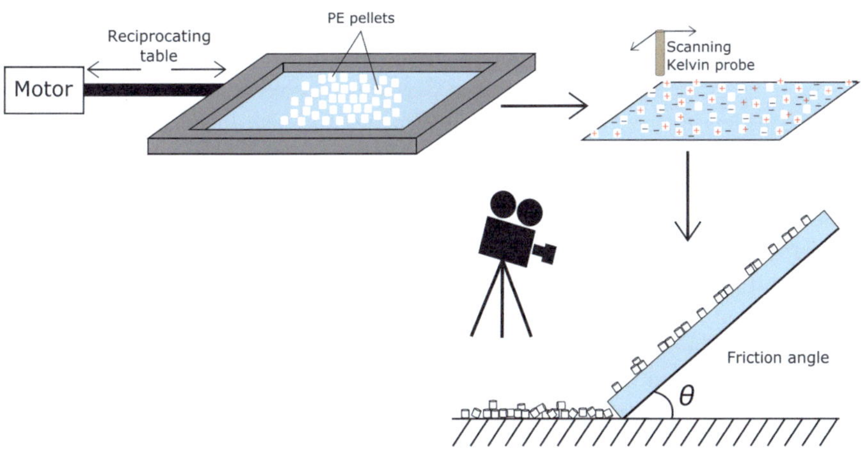

Fig. 8.5 Friction angle on tribocharged surfaces setup. Uncharged PTFE sheets framed on a thick aluminum sheet, carrying 30 PE pellets randomly placed on its surface were shaken for 5 min on a reciprocating table and then mounted on the swinging arm of the inclined plane, gradually raised in 5° steps and the number of pellets sliding at each angle was recorded

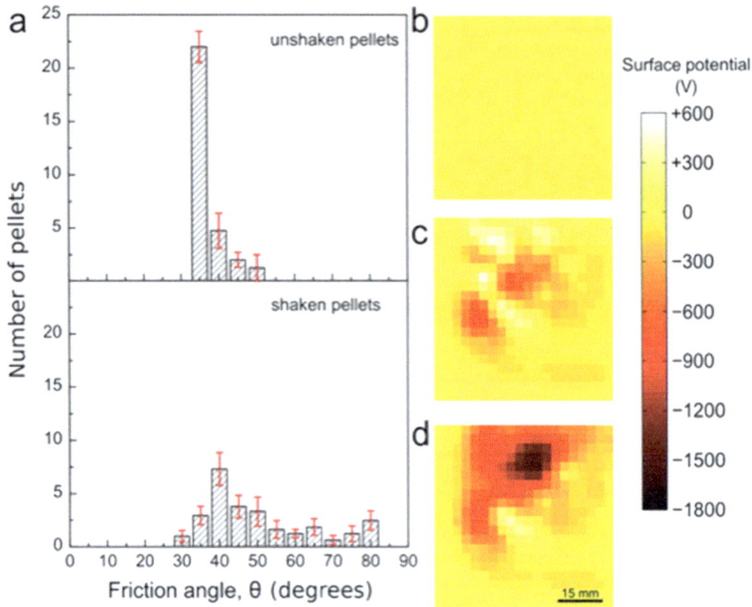

Fig. 8.6 Tribocharging effect on friction angles of PE pellets on PTFE. (**a**) Control measurement using unshaken pellets (*top*) and the distribution of values obtained by averaging the results of 13 shaking runs using 30 pellets each (*bottom*). Potential maps of: (**b**) PE pellets on clean PTFE prior to shaking, (**c**) pellets shaken for 300 s on PTFE and (**d**) PTFE after removal of PE pellets. Error bars are standard deviations of the average. Reprinted with permission from [31]

supporting the PTFE film and the pellets was mounted on the swinging arm of an inclined plane that was gradually raised in 5° steps, while recording the number of pellets sliding at each angle. Static friction coefficients $\mu_s = \tan \theta$ were then calculated and the histograms of pellet number vs. sliding friction angle are in Fig. 8.6. Friction angles of uncharged PE pellets over PTFE peaks at 40° but for the tribocharged pellets they spread toward much higher and also to lower angles (>90° down to 30°), after a short sliding time. This shows that tribocharging produces very large friction coefficients but some pellets have *lower* friction coefficients than the neutral ones.

In these experiments, the electrostatic measurements were done ex-situ, apart from the friction measurements. Thus, they do not give a real-time perspective on charge exchange and friction modification. Burgo and Erdemir [32] overcame this issue by doing in-situ experiments where charge exchange is measured online, concurrent with friction measurements. They used a rigorous setup in tribology, *the ball-on-disk* geometry, to record the macroscopic friction force while simultaneously measuring the current generated at the metal–insulator interface, as shown in Fig. 8.7. The results pointed out to a strong correlation between the macroscopic

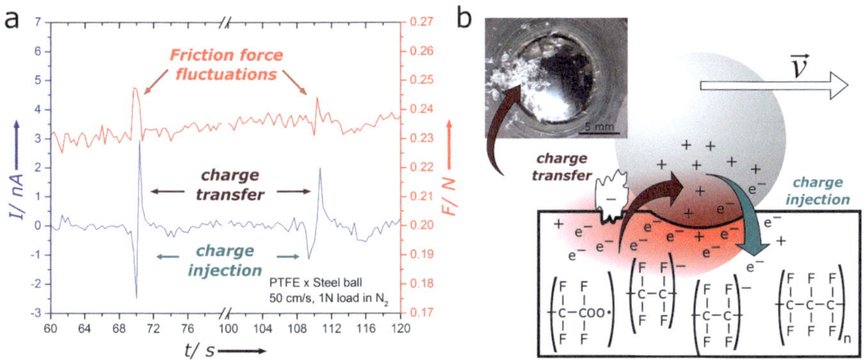

Fig. 8.7 Tribocurrent at metal–insulator interfaces. (**a**) Friction force fluctuations and tribocurrent generation at the metal–PTFE interface and (**b**) schematic representation of the underlying mechanism. Reprinted with permission from [32]

friction force and the electrostatic charge exchange at the interface. Figure 8.7a shows friction force and tribocurrent that were recorded simultaneously by sliding a steel ball on a PTFE surface under nitrogen atmosphere. The tribocurrent signal is always accompanied by a transient increase in the friction force signal, producing stick-slip behavior, as schematically described in Figure 8.7b.

Friction force fluctuations are often observed in tribological tests and Singer et al. have shown that this effect is generally caused by the presence of third bodies [15]. The micrograph in Fig. 8.7c was acquired after the tribological experiments and it shows macroscopic flake residues of PTFE adherent to the metallic surface, but only on the "*tail*" of the sphere, which means that the PTFE sticks to metal right after the ball slips on the surface. In fact, material transfer has also been pointed out as playing a key role in triboelectrification [23] and here we can state that friction force fluctuations are always followed by two tribocharging steps: first, an extra flow of electric charge from the ball to PTFE and then back. This can be understood recalling that during mechanical stress, electrons flow from compressed regions to extended ones, so that the Fermi level remains the same [33–35]. Consequently, extended regions have a negative charge density, when electrons are injected on PTFE, evidenced by the negative signal of the electric current. On the other hand, PTFE (which has a tendency to acquire negative potential in contact electrification) transfer negative charge back to the metal ball, mostly by material transfer and producing the positive current signal measured by the electrometer. Van der Waals and Coulombic interactions operate at the interfaces while chemical reactions also take place during the displacement of the metal ball on PTFE. Together, they change the contact area, reflecting directly on the macroscopic friction force.

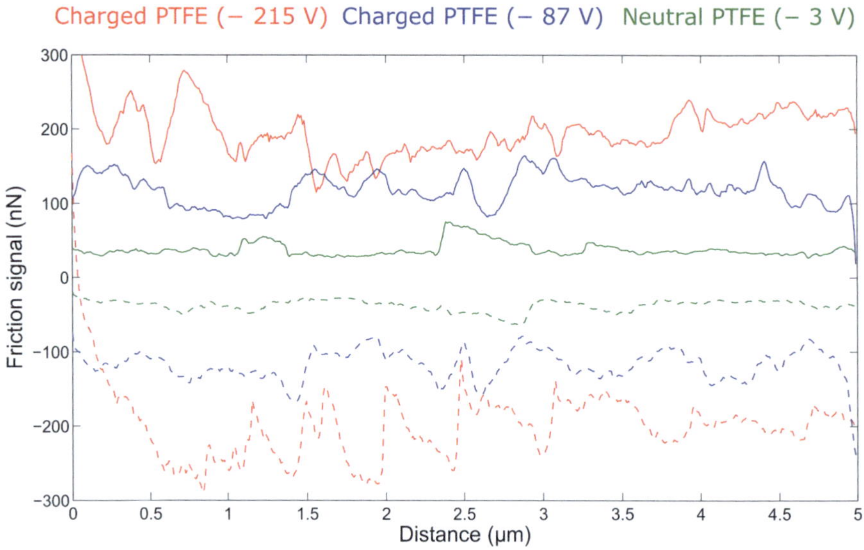

Fig. 8.8 Friction force signal on neutral and tribocharged PTFE. Friction signal profiles were extracted from FFM images. Reprinted with permission from [31]

8.3.2 AFM Experiments (LFM, Force-Distance and Nanomechanical Mode)

Tribocharge patterns are fractal [36] and they should thus display symmetry of scale. For this reason, the influence of electrostatic charge on friction should be observed at any scale, including the nanometric range. This assumption was experimentally verified using AFM in many modes and at least two different equipment configurations. Friction force microscopy (FFM) that is also called lateral force microscopy (LFM) has been invaluable in providing friction data with high spatial resolution. This technique is derived from contact mode imaging and it measures the lateral bending of the cantilever probe as it scans a surface. Results of FFM measurements on tribocharged surfaces are in Fig. 8.8, showing that the lateral force is close to zero on uncharged PTFE but it largely increases when the polymer surface is previously tribocharged. Compared to neutral PTFE, the friction signal increases roughly sevenfold when PTFE is tribocharged. A consequence of this result is that tribocharged PTFE is no longer a low surface energy material but it becomes a high surface energy solid, easily sticking to other materials.

Actually, the effect of electric charge on adhesive forces is better viewed with force vs. distance (*Fd*) curves and adhesion maps that are conveniently recorded using the PeakForce Quantitative Nanomechanical mode (PKQNM) [37]. Fd curves yield information on surface interactions, adhesion, elasto-plastic and many

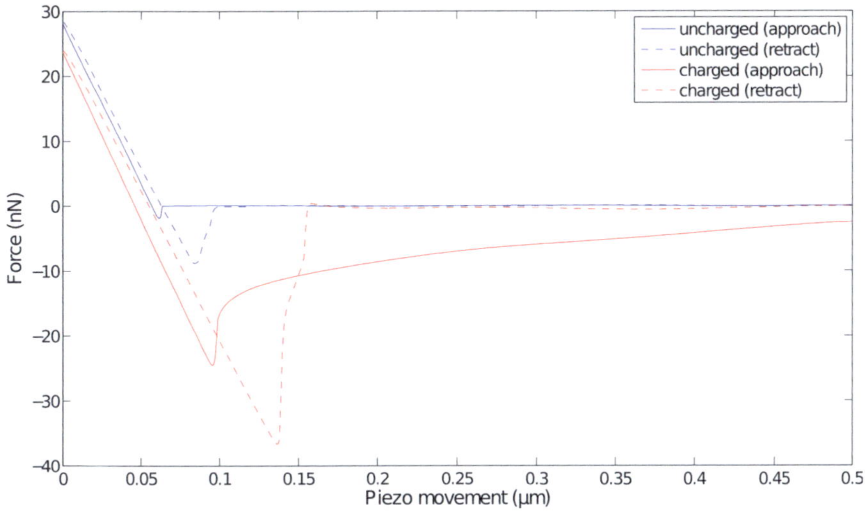

Fig. 8.9 Force–distance curves on neutral and tribocharged PTFE. Fd curves for approach and retraction of a silicon nitride tip from neat, uncharged PTFE and tribocharged PTFE. Average potential measured over the polymer with a macroscopic Kelvin electrode is −192 V. Reprinted with permission from [31]

other properties, which makes this a fundamental tool in surface science. Again, the results for neutral and previously tribocharged PTFE are markedly different. *Fd* curves [31] seen in Fig. 8.9 show that tip interaction with neutral PTFE is negligible until a short distance is attained and van der Waals attraction is observed. On the other hand, when PTFE is electrostatically charged, tip attraction during tip approach is observed at distances greater than 500 nm. Also, the pull-off force (or *adhesion force*) increase by at least a factor of 5, close to the change measured for friction signal.

The nanomechanical mode (PKQNM) has been successfully applied to the study of complex polymeric structures and relevant biological systems. PKQNM images are composed by acquiring a large number of Fd curves (commonly, 3 Fd curves per pixel) using the tapping mode in an AFM instrument. Thus, one of its capabilities is to provide adhesion maps and this was applied to explore surface properties of PTFE rubbed with a metal ball, as in the ball-on-disk tribology experiments. As seen in Fig. 8.10, the adhesive force on clean PTFE is less than 10 nN, but the adhesion on tribocharged PTFE reaches 150 nN in most pixels, due to electrostatic charge. This force range is the same as that observed for high-energy surface materials or even in geckos [38], explaining why charged PTFE strongly sticks to other materials [32].

Thus, Coulombic interactions due to surface charge may supersede all other contributions to macro and micro friction coefficient, at the macro and nanoscales.

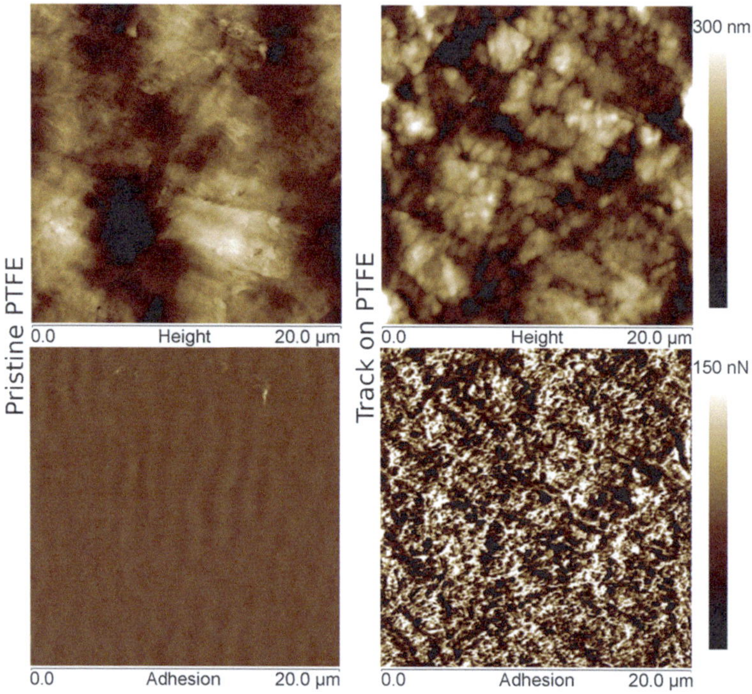

Fig. 8.10 Adhesion maps on neutral and rubbed PTFE. Topography (*top*) and adhesion (*bottom*) maps of pristine and tribotested (N_2, 2 N load, and 50 cm s^{-1}) PTFE obtained using the PeakForce Quantitative Nanomechanical mode (PKQNM). Reprinted with permission from [32]

This raises a new question: *How can the electrostatic charge be included in models for contact mechanics? Is it possible to add a Coulombic force contribution, simply?* To answer this question, we need to recall that details of contact electrification are still poorly known, in almost every dielectric material; there are even large differences between single contact, multiple contacts and sliding contact (rubbing) [35]. Also, when an insulator is part of the interface there is an inherent unpredictability [39], due to the chaotic nature of the micro [40] and macro [30] mosaics of charge developed during contact electrification. So, the answer to this question cannot be highly positive for every material, at the present time.

A Coulombic force contribution can be added in the cases of metals or semiconductors, when charge transferring follows work function properties and surface charge density is expected to be uniform. However, neither overall charge or charge density are more or less easily calculated for dielectrics and they can only be obtained from experimental measurements. So, the introduction of charging effects on contact mechanics experiments depends on previously acquiring the surface electrostatic maps (with macroscopic or microscopic Kelvin probes) and then using these samples in contact experiments, designed to minimize further charging or charge pattern modification. At this point, this seems to be the only way to proceed. Beyond, it is necessary to recall that most metal surfaces are coated with some oxide that is often a dielectric itself. So, even for metals strong fluctuations of charge density are to be expected.

8.4 Conclusions

There is a powerful interplay between friction and tribocharging: friction produces surface charge that in turn modifies friction coefficients that in turn affect friction forces. These mutual effects are mediated by chemical reactions added to mass transfer among the surfaces in contact. Moreover, tribocharge in dielectrics follows complex, fractal geometrical patterns that make its effects on friction quite complex. For this reason, they can only be assessed by well-designed experiments, in the absence of established theories. The mutual effects of friction and surface charge are certainly relevant to contact mechanics and materials wear but surface charge is not included in the relevant theories. Its inclusion is highly desirable but it will not be an easy task.

References

1. Bowden FP, Tabor D (1954) Friction and lubrication. Oxford University Press, Oxford
2. Hertz H, Reine J (1882) Ueber die Berührung fester elastischer Körper. Angew Math 92:156–171
3. Adamson AW, Gast AP (1997) Physical chemistry of surfaces. Wiley, New York
4. Johnson KL, Kendall K, Roberts AD (1971) Surface energy and the contact of elastic solids. Proc R Soc A Math Phys Eng Sci 324:301–313
5. Mate CM (2008) Tribology on the small scale: a bottom up approach to friction, lubrication, and wear. Oxford University Press, Oxford
6. Derjaguin BV, Muller VM, Toporov YUP (1975) Effect of contact deformation on the adhesion of particles. J Colloid Interf Sci 53:314–326
7. Tabor D (1977) Surface forces and surface interactions. J Colloid Interf Sci 58:2–13
8. Maugis D (1992) Adhesion of spheres: the JKR-DMT transition using a Dugdale model. J Colloid Interf Sci 150:243–269
9. Muller VM, Yushchenko VS, Derjaguin BV (1980) On the influence of molecular forces on the deformation of an elastic sphere and its sticking to a rigid plane. J Colloid Interf Sci 77:91–101
10. Shi X, Zhao Y-P (2004) Comparison of various adhesion contact theories and the influence of dimensionless load parameter. J Adhes Sci Technol 18(1):55–68
11. Chen YL, Israelachvili JN (1991) Molecular mechanisms associated with adhesion and contact angle hysteresis of monolayer surfaces. J Phys Chem 95:10736–10747
12. Yoshizawa H, Chen YL, Israelachvili J (1993) Fundamental mechanisms of interfacial friction. 1. Relation between adhesion and friction. J Phys Chem 97:4128–4140

13. Tabor D (1995) Tribology—the last 25 years. A personal view. Tribol Int 28(1):7–10
14. Mittal KL (1976) Adhesion aspects of metallization of organic polymer surfaces. J Vac Sci Technol 13:19–25
15. Singer IL, Dvorak SD, Wahl KJ, Scharf TW (2003) Role of third bodies in friction and wear of protective coatings. J Vac Sci Technol A 21:232–240
16. Park JY, Ogletree DF, Thiel PA, Salmeron M (2006) Electronic control of friction in silicon pn junctions. Science 313:186–186
17. Park JY, Qi Y, Ogletree DF, Thiel PA, Salmeron M (2007) Influence of carrier density on the friction properties of silicon pn junctions. Phys Rev B 76:064108
18. Altfeder I, Krim J (2012) Temperature dependence of nanoscale friction for Fe on YBCO. J Appl Phys 111:094916
19. Krim J (2012) Friction and energy dissipation mechanisms in adsorbed molecules and molecularly thin films. Adv Phys 61:155–323
20. Brezoczky B, Seki H (1990) Triboattraction: friction under negative load. Langmuir 6:1141–1145
21. Park JY, Salmeron M (2014) Fundamental aspects of energy dissipation in friction. Chem Rev 114:677–711
22. Burgo TAL, Rezende CA, Bertazzo S, Galembeck A, Galembeck F (2011) Electric potential decay on polyethylene: role of atmospheric water on electric charge build-up and dissipation. J Electrostat 69:401–409
23. McCarty LS, Whitesides GM (2008) Electrostatic charging due to separation of ions at interfaces: contact electrification of ionic electrets. Angew Chem Int Ed 47:2188–2207
24. Matta C, Eryilmaz OL, De Barros Bouchet MI, Erdemir A, Martin JM, Nakayama K (2009) On the possible role of triboplasma in friction and wear of diamond-like carbon films in hydrogen-containing environments. J Phys D Appl Phys 42:075307
25. Camara CG, Escobar JV, Hird JR, Putterman SJ (2008) Correlation between nanosecond X-ray flashes and stick–slip friction in peeling tape. Nature 455:1089–1092
26. Budakian R, Putterman SJ (2000) Correlation between charge transfer and stick-slip friction at a metal-insulator interface. Phys Rev Lett 85:1000–1003
27. Akbulut M, Godfrey Alig AR, Israelachvili J (2006) Triboelectrification between smooth metal surfaces coated with self-assembled monolayers (SAMs). J Phys Chem B 110(44): 22271–22278
28. Morris S, Wood RJK, Harvey TJ, Powrie HEG (2003) Electrostatic charge monitoring of unlubricated sliding wear of a bearing steel. Wear 255:430–443
29. Seto T (1995) Effects of an electric field on the static friction of a metal on a ferroelectric material. Appl Phys Lett 67:442–443
30. Burgo TAL, Ducati TRD, Francisco KR, Clinckspoor KJ, Galembeck F, Galembeck SE (2012) Triboelectricity: macroscopic charge patterns formed by self-arraying ions on polymer surfaces. Langmuir 28:7407–7416
31. Burgo TAL, Silva CA, Balestrin LBS, Galembeck F (2013) Friction coefficient dependence on electrostatic tribocharging. Sci Rep 3:2384
32. Burgo TAL, Erdemir A (2014) Bipolar tribocharging signal during friction force fluctuations at metal-insulator interfaces. Angew Chem Int Ed 53:12101–12105
33. Harper WR (1967) Contact and frictional electrification. Oxford at the Clarendon Press, London
34. Lowell J, RoseInnes AC (1980) Contact electrification. Adv Phys 29(9):947–1023
35. Lacks DJ, Sankaran RM (2011) Contact electrification of insulating materials. J Phys D Appl Phys 44(45):453001
36. Santos JP, Corpart P, Wong K, Galembeck F (2004) Heterogeneity in styrene- butadiene latex films. Langmuir 20:10576–10582
37. Pittenger B, Erina N et al (2012) Quantitative mechanical property mapping at the nanoscale with Peak Force QNM. Bruker Application Note #128, Bruker Nano Surfaces Divison Santa Barbara, USA

38. Geim AK, Dubonos SV, Grigorieva IV, Novoselov KS, Zhukov AA, Yu Shapoval S (2003) Microfabricated adhesive mimicking gecko foot-hair. Nat Mater 2(7):461–463
39. Lacks DJ (2012) The unpredictability of electrostatic charging. Angew Chem Int Ed 51:6822–6823
40. Baytekin HT, Patashinski AZ, Branicki M, Baytekin B, Soh S, Grzybowski BA (2011) The mosaic of surface charge in contact electrification. Science 333:308–312

Chapter 9
Electrostatic Adhesion

Contents

9.1 Contact Charging and Electrostatic Adhesion

Many years ago, Deryagin and colleagues presented and strongly defended the importance of electrostatic adhesion [1], largely based on the ubiquity of electrical double layer formation at interfaces, followed by Coulomb attraction. However, the conceptual spread and acceptance of electrostatic adhesion by researchers in the field of adhesion were impaired by disputes on the supporting experimental evidence that were probably largely due to well-known difficulties in making reproducible electrostatic measurements and to the lack of clear understanding of the mechanisms for insulator electrostatic charging.

The status of electrostatic adhesion can be assessed by examination of the widely used Handbook of Adhesion, by Packam [2]. It contains a short chapter [3] on electrical

F. Galembeck, T.A.L. Burgo, *Chemical Electrostatics*,
DOI 10.1007/978-3-319-52374-3_9

adhesion, discussing electrostatic precipitation of dust, photocopying machines, electrical effects on flowing powders and electro-gelation of fluids. In the handbook chapter on the Surface Forces Apparatus, the reader learns that "one of the few conclusive measurements indicating the validity of the electrostatic theory of adhesion" was described by Horn and Smith [4] that detected force of adhesion between sapphire and mica well in excess of the calculated work of adhesion, showing that charge had transferred between the two materials in contact. Finally, the chapter on Theories of Adhesion sums up the situation: there are some phenomena that cannot be explained without considering electrostatic forces, but the theories that had hitherto been advanced had been strongly criticized [5], leading many researchers studying adhesion and practitioners developing adhesives to belittle the importance of electrostatic contribution to adhesive joints.

Hays did a careful analysis, in his chapter within the "Fundamentals of Adhesion" book edited by Lee [6]: he examined the charging requirements for the onset of significant electrostatic adhesion while recognizing the difficulties to assess an essential parameter that is the surface charge density. Following his calculations, the electrical component of the adhesive force between planar surfaces of solids becomes important if the charge exchange density corresponds to 16,000 microcoulombs per square meter, or more, about 1% of all surface atoms. At this time, the maximum density measured upon separation of contacting materials was only one-tenth as much. However, charge density measured following surface separation is probably much lower than that existing during contact, that could not be measured directly.

More recently, there has been great progress in detecting electrostatic contributions to adhesion, in several systems, largely thanks to new experimental developments together with new models for charge build-up in solids. Great impact was made by the development of various implementations of scanning probe microscopies that provide maps of electric potential and electric charge on solid surfaces, with nanometric and even atomic resolution [7]. These receive various names and the most often used is Kelvin Force microscopy (KFM) [8] that revealed hitherto unsuspected electric patterns in every dielectric surface that was examined. Another development was the development and commercialization of practical Kelvin electrodes and electrometers. This topic is further developed ahead in this chapter and in Chap. 14 of this book, on Instrumentation.

The importance of electrostatic contribution to adhesion is now acknowledged in a number of important situations discussed in the following sections, from widespread technologies to biological systems. As for its role in adhesives, it competes with other various mechanisms (van der Waals forces, acid–base, chain entanglement, occlusion) that have greater or smaller importance in different systems.

9.2 Electrostatic Adhesion in Soft Materials

Even though the importance assigned to electrostatic attraction in adhesion has varied with time and authors, it plays a decisive role in many technologies or technological problems that are discussed in Chap. 13.

It also has paramount importance in biological systems making a decisive contribution to the stability of complex conformation in proteins, ribosomes and other important biological materials that are not discussed in this book.

9.2.1 Rubber and Other Latexes

Both natural and synthetic polymers are widely used in the formulation of adhesives and coatings, often compounded with many other materials. In a related context, different polymers are combined with other polymers to make polymer blends or with glass, ceramic, various powders and metals to make polymer composites. Interest in combining polymers with other materials has grown continuously due to the increasing demand for more diverse and complex functional materials displaying the desired combinations of electrical, mechanical, chemical, barrier, thermal and optical properties. This requires the control of interfacial tension and rheology between the diverse constituents of multiphase particles, fibers and films, to achieve the desired practical adhesion [9]. In many cases, this is achieved by creating the conditions for the establishment of electrostatic adhesion even though this is hardly acknowledged in current literature [10]. This negligence in intriguing, because the simple existence of an already causes charge separation [11].

9.2.2 Layer-by-Layer Fabrication

A remarkable case of application of electrostatic interactions in materials fabrication and modification is the "layer-by-layer" nanostructure formation that has produced outstanding results, especially in biological systems [12]. Schematically, this works by immerging a substrate within solutions of oppositely charged polyelectrolytes and annealing the adsorbed layer. As a result, layers of alternate charge are obtained forming a multilayer or laminate macrocrystal that can be used as nanoparticle carrier. For instance, nanotubes were fabricated through the layer-by-layer (LbL) assembly of poly-L-lysine hydrochloride (PLL), poly-L-glutamic acid (PGA) and magnetic nanoparticles within the pores of polycarbonate templates with subsequent removal of the template [13]. The method was applied to nanoparticle coating [14] as a means of generating drug-releasing surfaces for biomedical applications, from small molecules to biological drugs and nucleic acids [15].

This is now sufficiently developed to be present at the basis of some technological applications [16].

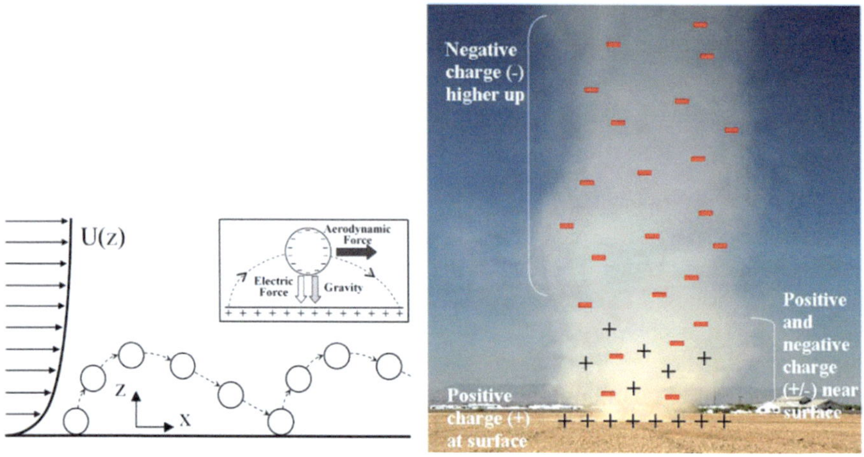

Fig. 9.1 *Left*: Schematic representation of saltation, showing the logarithmic wind profile U (z) to the left of a sand particle bouncing along the surface. Inset: charge distribution in the particle over the positively charged soil surface. *Right*: Hypothesized charge distribution in a dust cloud (i.e., a dust devil). The small dust particles move upward through convection or turbulent diffusion, while the heavier saltating particles stay closer to the surface producing large electric fields. Reprinted with permission from [17]

9.2.3 "Saltation" and Dust Adhesion

"Saltation" is used for some authors to refer to wind-blown sand. This is an important geological process, as the primary source of atmospheric mineral dust aerosols. Classical saltation theory neglects the electrostatic charging of particles and it shows significant discrepancies from measurements, but these discrepancies can be resolved by the inclusion of sand electrification in a physically based saltation model (Fig. 9.1) [17, 18]. Following this work, electric forces enhance the concentration of saltating particles and cause them to travel closer to the surface, in agreement with measurements.

Moreover, these results show that sand and other aerosol particles often carry charge that contributes electrostatic interactions to adhesion with photovoltaic glass panels [19] and other surfaces, largely impairing their performance on Earth and also in space research [20]. This also helps to understand the interest in electrostatic procedures [21] for particle removal and surface cleaning.

9.3 Microchemical Evidence for Electrostatic Adhesion in Materials

Experimental limitations for assessing double layer and other charge distribution patterns in solids were alleviated following the introduction of new microscopy techniques and progress in instrumentation for electrostatic measurements. Scanning probe techniques were introduced, based on the atomic force microscopy (AFM), especially the Kelvin force microscopy (KFM) and electric force microscopy (EFM) and they became commercially available in the mid-90s, producing much new and then unexpected information on the distribution of electric charges and potentials across insulator surfaces [22, 23]. Moreover, micromanipulation techniques have contributed to recognition of the importance of electrostatic adhesion that was evidenced by detailed mathematical modeling [24].

On the other hand, the increasing availability of electron spectroscopy imaging (ESI) based on energy-filtered transmission electron microscopy (EFTEM) has produced detailed elemental maps and it is now producing molecular maps of soft materials like polymer blends and nanocomposites, showing how different constituents of a sample are distributed in its interior and thus providing much unprecedented microchemical information with high spatial resolution [25, 26]. The combination of scanning electric probe techniques and EFTEM produced detailed pictures of ionic component distribution in polymer-clay bonds [27–31].

Microscopy techniques were invaluable in revealing the complexity of natural rubber nanostructure [32]. Natural rubber contains a number of inorganic and organic components beyond the base poly(cis-1,4-isoprene) and these were found to account for its still unmatched properties, due to different mechanisms. First, the rubber frequently contains nanosized particles formed by many elements: Al, Ca, S, P, O and others (Fig. 9.2) [33]. Moreover, calcium is often found associated with gel domains inside rubber particles in the latex and later in the dry rubber, evidencing that it has a cross-linking role based on electrostatic interactions [34].

The participation of electrostatic adhesion is essential in rubber-clay nanocomposites prepared by mixing rubber latex and aqueous sodium bentonite and sodium montmorillonite [35, 36]. This type of procedure is now known in the literature as the latex route for nanocomposite preparation and it has some advantages over the well-established techniques that require clay surface modification prior to mixing with the polymer. Nanocomposites of natural rubber latex and layered silicates can be prepared by a mild dispersion shear blending process and the results of X-ray diffraction (XRD, in Fig. 9.3) and transmission electron microscopy (TEM, in Fig. 9.4) show that clay particles are well dispersed in the dry latex where the platelets have a preferential orientation, forming translucent nanocomposites [37]. Moreover, contact between clay platelets and rubber is flawless and this is a strong evidence for strong adhesion. Otherwise, the two phases would be separated when they were sliced.

The nanocomposites show tensile mechanical (Fig. 9.5) properties analogous to those obtained with vulcanized rubber as well as an increased solvent resistance, which is expected considering that there is significant adhesion between clay lamellae and rubber. Nanocomposite swelling is anisotropic. Natural rubber properties are thus strongly modified by nanocomposite formation producing unprecedented combinations of properties.

The adhesion between rubber and clay is also evidenced in the analytical TEM images in Fig. 9.6, acquired from a sample obtained by drying a dilute dispersion containing clay and rubber latex particles. There is strong superimposition between

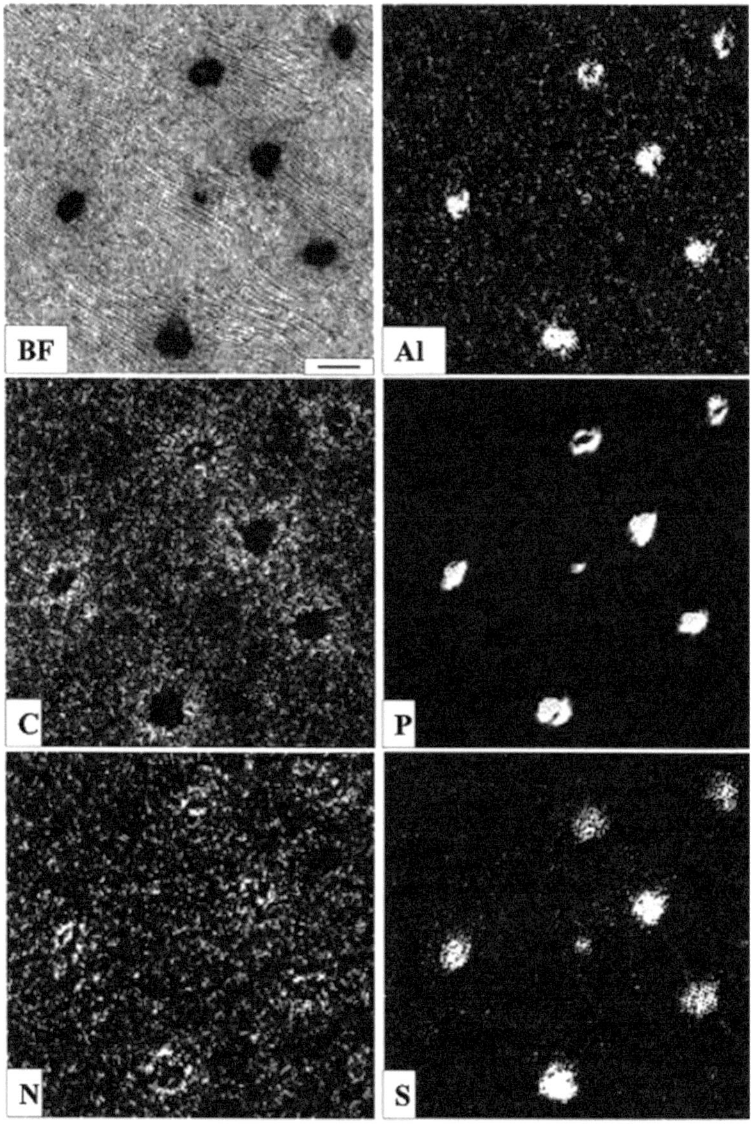

Fig. 9.2 Bright-field and elemental images from dialyzed rubber latex film, average thickness 63 nm. The dark spots in the bright-field image are inorganic particles containing Al, P, S. C map shows that the polymer adheres strongly to these particles. Scale bar is 150 nm [33]

the clay particles (that are evidenced morphologically and also in the Si elemental map) and rubber that is also evidenced by the elemental C map.

Key adhesion mechanisms in this process are capillary adhesion during dispersion drying followed by electrostatic adhesion mediated by counter-ions, schemati-

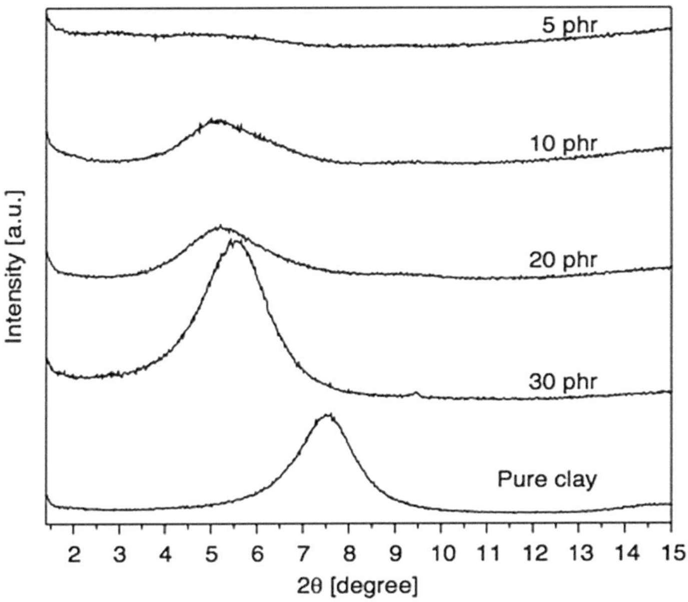

Fig. 9.3 X-ray diffractograms of natural rubber-montmorillonite nanocomposites. Clay concentrations are in per hundred ratio (phr). The diffraction line shown for pure clay corresponds to the distance between lamellae that increases due to polymer insertion within lamellae stacks or tactoids. Reprinted with permission from [37]

cally described in Fig. 9.7, for the formation of nanocomposite films of Stöber silica and an acrylic latex [38].

This has an interesting consequence: the mechanical properties of the nanocomposites are strongly dependent on the intervening counter-ions [39] that are found in the clay or in the latex used as raw materials. This means the nanocomposite properties can be tuned by changing the counter-ions, which are minor and chemically inert components of the material. Moreover, the removal of hydration water strengthens the nanocomposite, shown in Fig. 9.8. As expected, the effect is most pronounced in the case of Li^+ ions that hydrate more extensively than sodium or potassium.

A schematic description of latex-clay nanocomposite formation including the removal of the hydration shell is in Fig. 9.9.

The adhesion of synthetic latex to aluminum polyphosphate nanosized particles and aggregates is another case of strong adhesion that was attributed to electrostatic adhesion [40].

A generalization is then possible: when particles from different phases dispersed in water are wettable and suitable counter-ions are present, dispersion drying produces strong capillary adhesion first, followed by electrostatic adhesion that makes strong contribution to materials cohesion and bond stability.

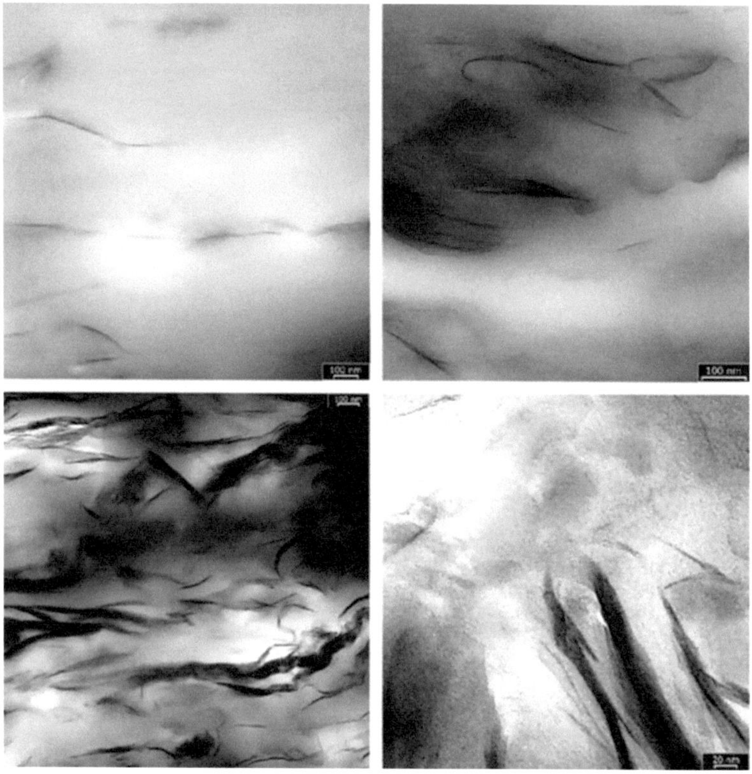

Fig. 9.4 TEM bright field micrographs from 5 (top) to 30 phr (bottom) thin cuts of rubber-montmorillonite nanocomposites. The cut were made normal to the nanocomposite film plane showing that lamellae are roughly aligned along the plane. Reprinted with permission from [37]

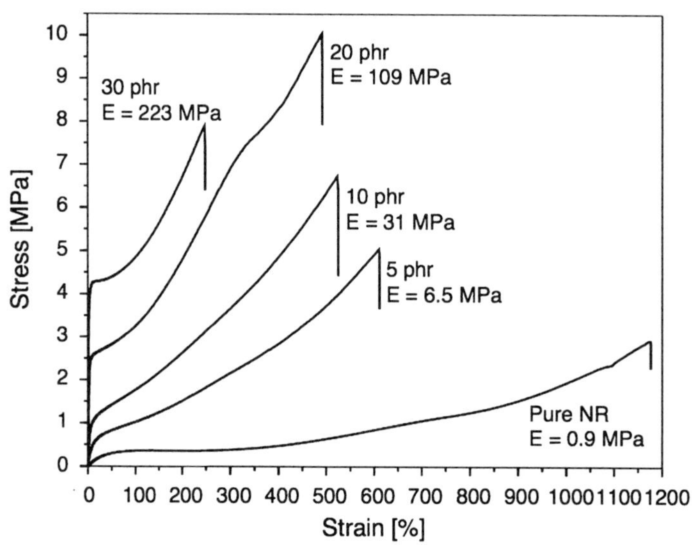

Fig. 9.5 The strength of natural rubber-montmorillonite nanocomposites. Changing the clay concentration (given in phr, per hundred ratio) produces large increase in the initial modulus, while increasing the energy required for breaking the material. Reprinted with permission from [37]

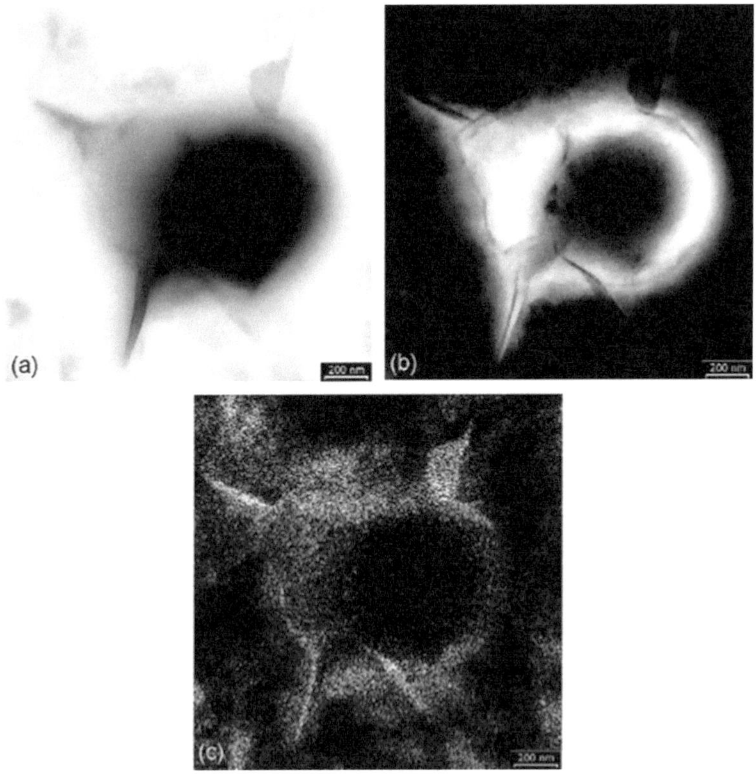

Fig. 9.6 Transmission electron micrographs from samples prepared by drying droplets of rubber-clay aqueous dispersion on top of thin carbon films: (**a**) bright-field image; (**b**) carbon map obtained by energy-filtered TEM; (**c**) silicon map. In the maps, the bright areas show the presence of the mapped element. Reprinted with permission from [37]

One earlier result of the application of microscopy techniques to elucidate bond formation was in the case of natural rubber adhesion to float glass sheets. Rubber films formed by casting latex on glass were easily peeled off, while films prepared with latex modified with sodium polyphosphate were strongly adherent [41]. KFM images from fracture surfaces (Fig. 9.10) show a charge sandwich at the rubber–glass interface that was interpreted as the outcome of colloidal phase separation in the latex, forming large mobile positive and negative charge domains. The glass negative surface was covered with a thin layer of positive domains that contribute to glass–rubber adhesion, adding to other ubiquitous factors like van der Waals and hydrogen-bond interactions.

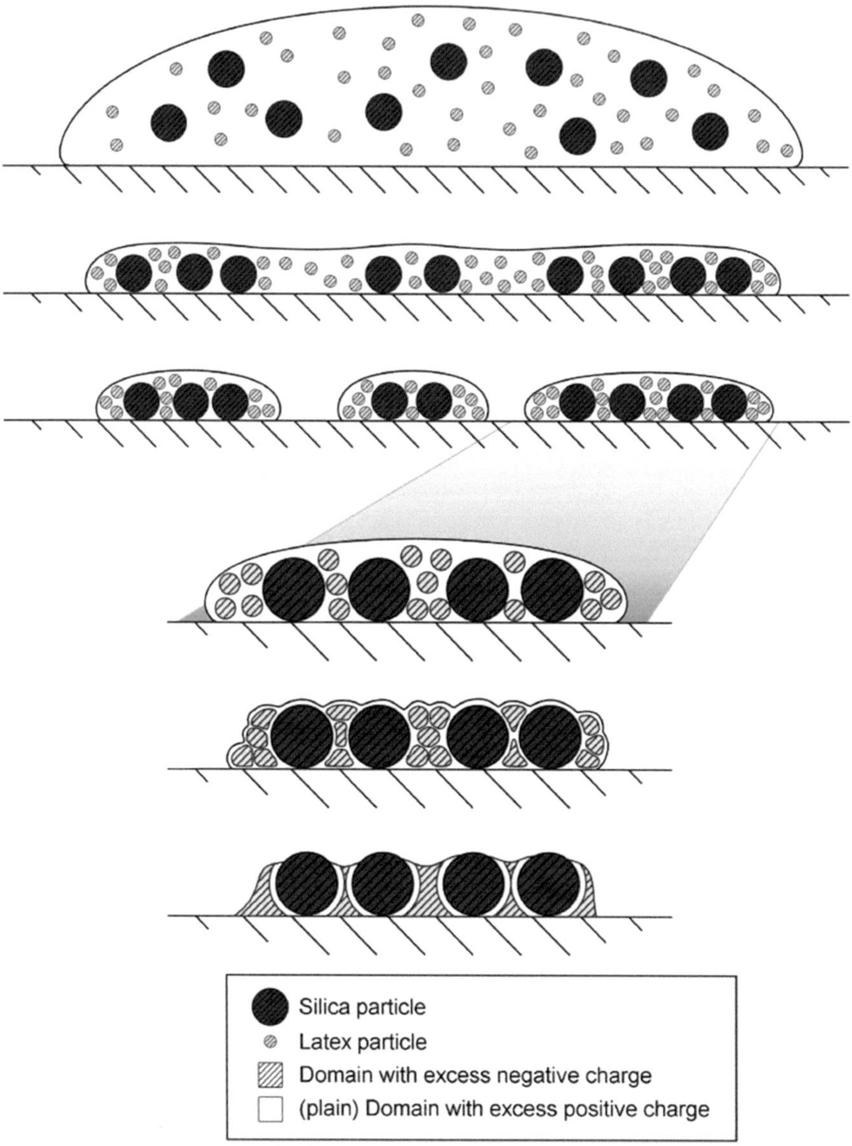

Fig. 9.7 Schematic description of the successive events involved in the adhesion of Stöber silica to styrene-acrylic latex, forming a nanocomposite film. Reprinted with permission from [38]

9.4 Theoretical Estimates

Model calculations have shown that electrostatic interaction energy in the clay-rubber bonds is in the 105 kJ/mol range, comparable to covalent bonds [39]. Mechanical energy for elastic deformation of the nanocomposites was determined

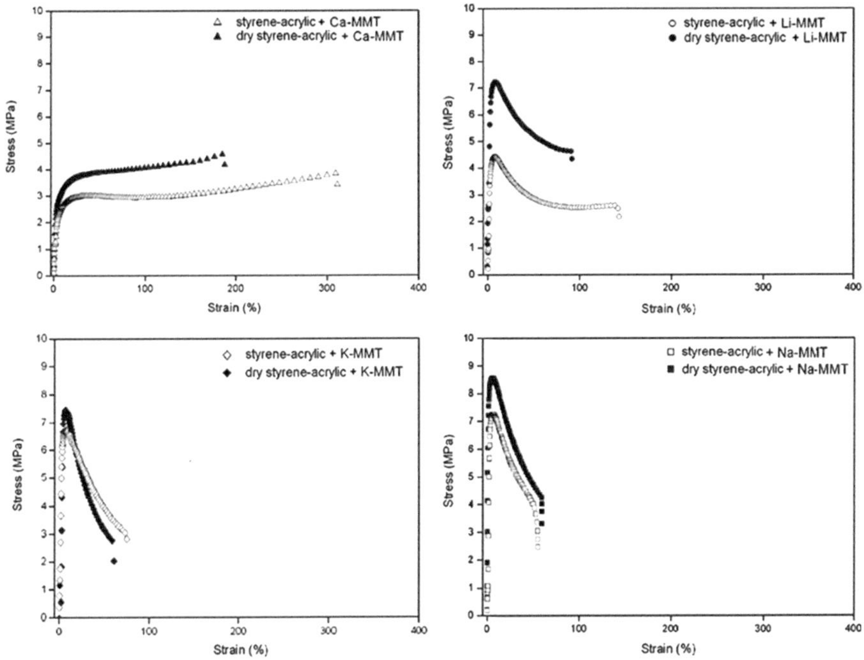

Fig. 9.8 Drying effect on the mechanical properties of clay-acrylic nanocomposites prepared with ion-exchanged clays. Each pair of plots allows the comparison of air-dried (25 °C) and oven-dried (120 °C) nanocomposite films. Reprinted with permission from [39]

experimentally and Table 9.1 shows that the measured values agree with the values calculated using an electrostatic model, thus confirming the decisive role of electrostatic interactions in the properties of these materials. An interesting outcome of these calculations is that stable electrostatic bonds may not be electroneutral, since the minimum free energy is obtained when the overall charge summation is non-zero. This is an interesting case for the spontaneous onset of excess charge during the approximation to thermodynamic equilibrium, a topic that was extensively discussed in early chapters.

9.4.1 Gecko Adhesion

Geckos walk on walls and hang from ceilings with the help of setae, hierarchical fibrils on their toe pads. Recently, they inspired the design and fabrication of fibrillar dry adhesives. The unique fibrillar feature of the toe pads of geckos allows them to develop an intimate contact with the substrate when the toe setae have the possibility to exchange significant numbers of electric charges with the substrate, producing contact electrification. Electrostatic contribution to the dry adhesion of geckos has

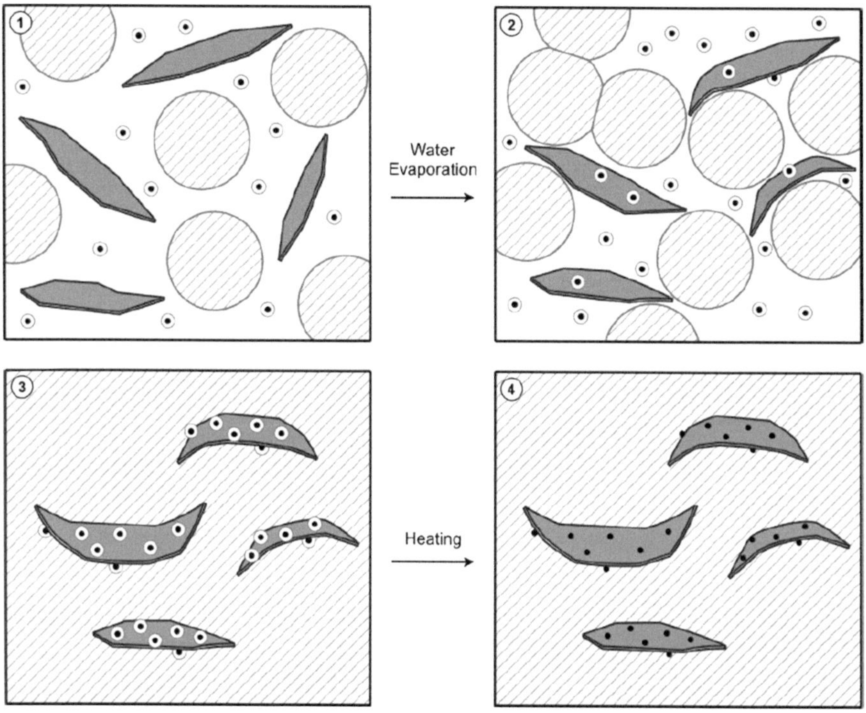

Fig. 9.9 Schematic representation of the role of positive counter-ions (*small circles*) on the adhesion between clay lamellae and polymer. (**a**) Latex and clay dispersion, (**b**) polymer—clay particles approach during dispersion drying, (**c**) hydrated counter ions bridging clay and the dry polymer matrix, and (**d**) dehydrated counter ions bridge clay lamellae and the polymer matrix. Reprinted with permission from [39]

not been considered in the literature but recent measurements of the magnitude of the electric charges and adhesion forces showed that contact electrification contributes to gecko adhesion, superseding the van der Waals or capillary forces which are conventionally considered as the main source of gecko adhesion [42].

9.4.2 Bacterial and Cell Adhesion

Bacterial adhesion plays an important role in the strategies of any living being and mammalian cells contain a complex apparatus that is used to establish or to evade adhesion. The initial stages of bacterial adhesion to a model-surface of sulfonated polystyrene can be described using hydrophobic and electrostatic parameters and further work was done using different model surfaces: (a) glass, as a model for hydrophilic and natural surfaces of silicates and oxides, (b) polystyrene coated with proteins, as a model for a surface coated with an organic layer, and (c) river Rhine

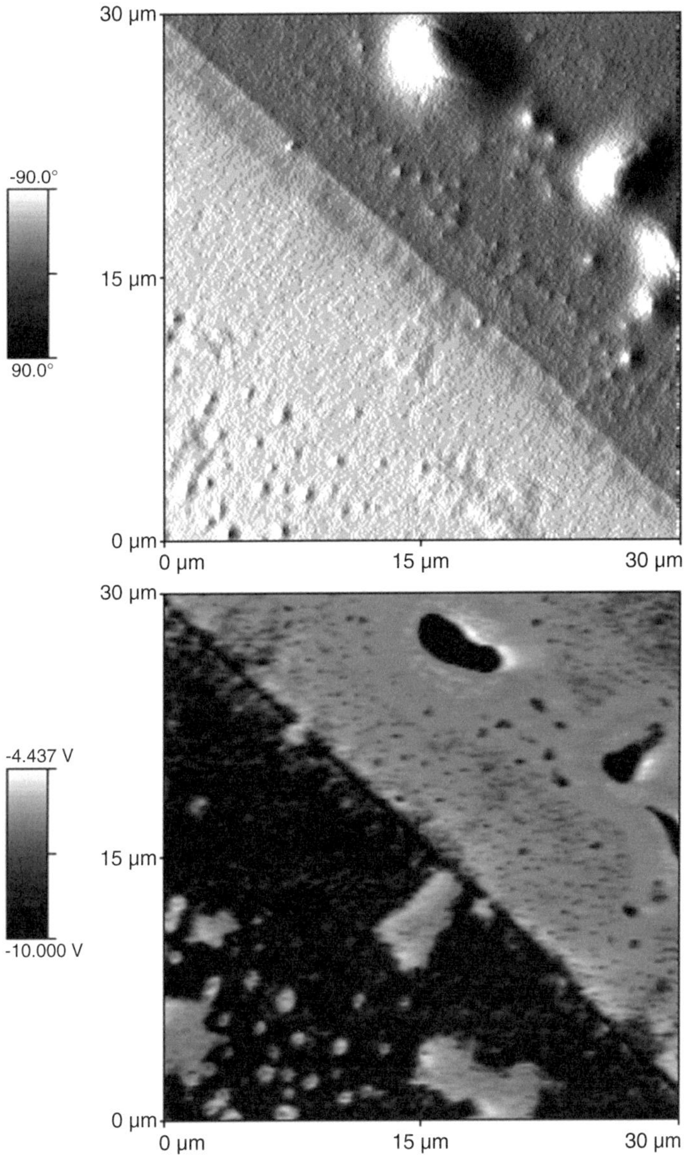

Fig. 9.10 AFM (*top*) and scanning electron potential microscopy (SEPM) bottom images from the NRLPP film/glass joint fracture surface. Glass is at the left lower side and the rubber-polyphosphate composite is at the right upper side. In the lower micrograph, notice the brighter domains close to the glass surface. Reprinted with permission from [41]

sediment, as an example of a natural surface. Electrostatic interactions dominate adhesion to glass, while adhesion to polystyrene coated with various types of proteins depends on protein type and the surface characteristics of the bacteria. The authors related van der Waals interactions to hydrophobicity of the interacting

Table 9.1 Elastic deformation work for rubber-clay nanocomposites. Reprinted with permission from [38]

	Work for 1% deformation (J)	Difference between nanocomposite and rubber (J)	Moles of ions in the test sample	Elastic energy per mole of ions (J/mol)
Rubber	0.386×10^{-3}	–	–	–
Rubber – Li^+-MMT	2.17×10^{-3}	1.78×10^{-3}	2.43×10^{-5}	73.7
Rubber – Na^+-MMT	4.97×10^{-3}	4.58×10^{-3}	4.91×10^{-5}	93.4
Rubber – K^+-MMT	4.66×10^{-3}	4.27×10^{-3}	2.99×10^{-5}	143

species, interpreting the adhesion of bacteria to the three types of surfaces used, including the river Rhine sediments, with the help of the DLVO-theory that is widely used in the study of colloidal stability. Thus, the conceptual principles of the DLVO-theory are suitable to describe, at least qualitatively, the initial processes of bacterial adhesion to a wide range of surfaces [43].

9.4.3 Biomolecules

Electrostatics plays a major role in determining the tertiary structure and the function of biomolecules. For this reason, it is a major focus of experimental, theoretical and computational studies of macromolecules, including the effects of the solvent and ions surrounding the biomolecule. There are many coexisting computational approaches that have been critically examined, from time to time [44].

9.5 Electroadhesion

A biased electrode inducing opposite charge on an adjacent surface will adhere to it. Since this depends on the electrical field but not on any electrical current, the electrode may be blocked, this means, separated from the surface by a compliant insulator that self-accommodate to the roughness of the neighbor surface. This was initially used for robotic grippers, in the microelectronics industry and in space application later extending to other fields. Electroadhesion will be discussed in detail in Chap. 13, on Electrostatic products and processes. However, it should be mentioned here that its performance may depend on ambient humidity, as indicated in a recent report from NASA and SRI International [45]. The combination of electroadhesion and electrostatic actuation was recently demonstrated [46].

9.6 A Valuable Tool for "Green" Fabrication

Electrostatic adhesion is now acknowledged as a valuable tool in the making of soft materials and it is especially well suited for soft materials derived from biomass.

This chapter showed how electrostatic adhesion is important in water-based processes performed at room temperature or in its vicinity, at ambient pressure. Recognizing its importance, we can probably improve material properties considering some desirable applications.

One interesting possibility is to achieve better understanding of the stability and durability of adobe used for construction, made with the soil or mud available in different places [47]. Making adobe has many features in common with the latex process for nanocomposites that was mentioned in this chapter. It is not unlikely that this can lead to safer use of adobe, alleviating building costs in many places. The same argument applies to water-based coatings and adhesives.

References

1. Deryagin BV, Krotova et al (1978) Adhesion of solids. Springer, New York
2. Packham DE (2005) Handbook of adhesion. Wiley
3. Kendall K (2005) Electrical adhesion. In: Packham DE (ed) Handbook of adhesion. Wiley
4. Horn RG, Smith DT (1992) Contact electrification and adhesion between dissimilar materials. Science 256(5055):362–364
5. Allen KW (2005) Diffusion theory of adhesion. In: Packham DE (ed) Handbook of adhesion. Wiley
6. Hays DA (1991) Role of electrostatic. In: Lee LH (ed) Fundamentals of adhesion. Springer, New York, p 249–278, chapter 8
7. Noy A, Frisbie CD et al (1995) Chemical force microscopy: exploiting chemically-modified tips to quantify adhesion, friction, and functional group distributions in molecular assemblies. J Am Chem Soc 117(30):7943–7951
8. Galembeck F (2002) Interfacial behavior of latex dispersions and emulsions. In: Hubbard AT (ed) Encyclopedia of colloid and surface science, vol 1. Marcel Dekker, pp 2677–2690
9. Mittal KL (1980) Interfacial chemistry and adhesion-recent developments and prospects. Pure Appl Chem 52(5):1295–1305
10. Park SJ, Seo, MK (eds) (2011) Solid-solid interfaces. Interface science and technology, vol. 18. Academic, p 252–331, chapter 4
11. Adamson AW, Gast AP (1987) Physical chemistry of surfaces. Wiley
12. Ladam G, Schaaf P et al (2002) Protein adsorption onto auto-assembled polyelectrolyte films. Biomol Eng 19:273–280
13. He Q, Tian Y et al (2008) Layer-by-layer assembly of magnetic polypeptide nanotubes as a DNA carrier. J Mater Chem 18:748–754
14. Schneider G, Decher G (2008) Functional core/shell nanoparticles via layer-by-layer assembly, investigation of experimental parameters for controlling particle aggregation and for enhancing dispersion stability. Langmuir 24:1778–1789
15. Hammond PT (2012) Building biomaterial materials layer-by-layer. Mater Today 15(5):196–206
16. Ettrich C, Schwamb M et al (2016) Method for coating surfaces and use of the coatings produced by this method. US Patent 20160136685A1, 19 May 2016

17. Renno NO, Kok JF (2008) Electrical activity and dust lifting on earth, mars, and beyond. Space Sci Rev 137:419–434
18. Kok JF, Renno NO et al (2000) Electrostatics in wind-blown sand. Phys Rev Lett 100:14501
19. Dastoori K, Al-Shabaan G et al (2016) Impact of accumulated dust particles'charge on the photovoltaic module performance. J Electrostat 79:20–24
20. Walton OR Adhesion of lunar dust. Nasa Technical Reports Server (NTRS). http://ntrs.nasa.gov/search.jsp?R=20070020448. Accessed 01 Jul 2016
21. Calle CI, Buhler CR et al (2009) Particle removal by electrostatic and dielectrostatic forces for dust control during lunar exploration missions. J Electrostat 67:89–92
22. Drelich J, Mittal KL (eds) (2005) Atomic force microscopy in adhesion studies. VSP/Brill, Leiden
23. Galembeck A, Costa CAR et al (2001) Scanning electric potential microscopy of polymers: electrical charge distribuition in dielectrics. Polymer 42:4845–4851
24. Lhernould MS, Berke P et al (2009) Variation of the electrostatic adhesion force on a rough surface due to the deformation of roughness asperities during micromanipulation of a spherical rigid body. J Adhes Sci Technol 23:1303–1325
25. Libera MR, Egerton RF (2010) Advances in the transmission electron microscopy of polymers. Polym Rev 36:321–339
26. Linares EM, Leite CAP et al (2009) Molecular mapping by low-energy-loss energy-filtered transmission electron microscopy imaging. Anal Chem 81:2317–2324
27. Rippel MM, Lee LT et al (2003) Skim and cream natural rubber particles: colloidal properties, coalescence and film formation. J Colloid Interface Sci 268:330–340
28. Braga M, Costa CAR et al (2001) Scanning electric potential microscopy imaging of polymer latex films: detection of supramolecular domains with nonuniform electrical characteristics. J Phys Chem B 105:3005–3011
29. Santos JP, Corpart P et al (2004) Heterogeneity in styrene-butadiene latex films. Langmuir 20:10576–10582
30. Teixeira-Neto E, Kaupp G et al (2004) Spatial distribuition of serum solutes on dry latex submonolayers determined by SPEPM, SNOM and SC microscopy. Colloid Surf A 243:79–87
31. Teixeira-Neto E, Kaupp G et al (2003) Latex particle heterogeneity and clustering in films. J Phys Chem B 107:14255–14260
32. Rippel MM, Leite CAP et al (2005) Formation of calcium crystallites in dry natural rubber particles. J Colloid Interface Sci 288:449–456
33. Rippel MM, Leite CAP, Galembeck F (2002) Elemental mapping in natural rubber latex films by electron energy loss spectroscopy associated with transmission electron microscopy. Anal Chem 74:2541–2548
34. Rippel MM, Leite CAP et al (2005) Direct imaging and elemental mapping of microgels in natural rubber particles. Colloid Polym Sci 283:570–574
35. Varghese S, Karger-Kocsis J (2003) Natural rubber-based nanocomposites by latex compounding with layered silicates. Polymer 44:4921–4927
36. Wu YP, Wang YQ et al (2005) Rubber-pristine clay nanocomposites prepared by co-coagulating rubber látex and clay aqueous suspension. Compos Sci Technol 65:1195–1202
37. Valadares LF, Leite CAP, Galembeck F (2006) Preparation of natural rubber-montmorillonite nanocomposite in aqueous medium: evidence for polymer-platelet adhesion. Polymer 47:672–678
38. Valadares LF, Linhares EM et al (2008) Electrostatic adhesion of nanosized particles: the cohesive role of water. J Phys Chem C 112:8534–8544
39. Bragança FC, Valadares LF et al (2007) Counterion effect on the morphological and mechanical properties of polymer-clay nanocomposites prepared in an aqueous medium. Chem Mater 19(13):3334–3342
40. Silva CA, Galembeck F (2010) Void formation and opacity in latex films with aluminum phosphate particles. Colloid Surf A 363:146–154
41. Rippel MM, Costa CAR et al (2004) Natural rubber latex modification by sodium polyphosphate: a SPM study on the improved latex adhesion to glass sheet. Polymer 45:3367–3375

42. Izadi H, Stewart KME et al (2014) Role of contact electrification and electrostatic interactions in gecko adhesion. J R Soc Interface 11:20140371
43. van Loosdrecht MCM, Norde W, Lyklema J, Zehnder AJB (1990) Hydrophobic and electrostatic parameters in bacterial adhesion. Aquat Sci 52(1):103–114
44. Koehl P (2006) Electrostatics calculations: latest methodological advances. Curr Opin Struct Biol 16:142–151
45. Bryan T, Macleod T et al (2015) Innovative electrostatic adhesion technologies. NASA Marshall Space Flight Center. http://ntrs.nasa.gov/archive/nasa/casi.ntrs.nasa.gov/20150021399.pdf. Accessed 15 Aug 2016
46. Shintake J, Rosset S et al (2016) Versatile soft grippers with intrinsic electroadhesion based on multifunctional polymer actuators. Adv Mater 28:231–238
47. Lertwattanaruk P, Choksiriwanna J (2011) The physical and thermal properties of adobe brick containing bagasse for earth construction. Built 1:53–62

Chapter 10
Self-assembly

Contents

10.1 From Disorder to Order

Science and engineering students are often led to believe that equilibrium is reached when the system under study reaches maximum entropy and thus it is maximally disordered. Accordingly, any material transformation of a system should bring it closer to chaos, in the sense of maximal disorder.

Students of chemistry, chemical engineering, biochemistry and related disciplines are vaccinated against the views expressed in the previous paragraph. This is because they soon learn that minimal Gibbs energy is the correct criterion for spontaneity of material transformations under the relevant conditions in our environment: constant P and T or better, contact with a pressure and temperature reservoir, like the surrounding atmosphere. Recognizing this spares researchers and engineers from endless and fruitless discussions on the final destination of matter.

Biologists are still readier to accept that ordering may be the spontaneous outcome of natural processes because their objects of study are more and more complex

F. Galembeck, T. A. L. Burgo, *Chemical Electrostatics*,
DOI 10.1007/978-3-319-52374-3_10

Fig. 10.1 Various forms of Liesegang rings at the Motoyama outcrop. Reprinted with permission from [4]

structures that have been formed from common matter, by evolutionary processes through ages.

Turing [1], Prigogine [2] and many other scientists made important contributions showing how orderly structures are formed during spontaneous processes, this means: *dissipative structures* are formed spontaneously in thermodynamic systems far from equilibrium. For instance, the Liesegang rings that are concentric rings formed when a precipitation reaction takes place simultaneously with reactant diffusion, within a system approaching equilibrium [3]. Structures that may have been made by processes like those modeled by Turing are frozen in beautiful mineralogical specimens as in Fig. 10.1 [4]. The formation of complex structures can also be modeled by cellular automata [5] that have an interesting feature: some automata grow to a point and stop growing at some definite step, in a way reminiscent of living beings.

The creation of more and more elaborate and/or well-defined material structures in the laboratory is a natural step for the chemists that progressed through chemical synthesis to supramolecular and macromolecular structures, now moving toward chemical biology. It is also natural within materials laboratories that made high-purified silicon, followed by complex heterostructures and more recently self-supporting semiconductor membranes [6], by design.

Self-assembly is the autonomous organization of components into patterns or structures and self-assembling processes are common throughout nature and technology [7]. They take place in the absence of direct human intervention, within many scales: from molecular and ionic components forming crystals to planetary-scale processes in the atmosphere and oceans. Much of the work in self-assembly is done on molecular systems but its attractiveness to nano- and micro-fabrication is obvious. For this reason, many of the most interesting applications of self-assembling processes are found at larger sizes, from nanometers to micrometers (Fig. 10.2) [8,

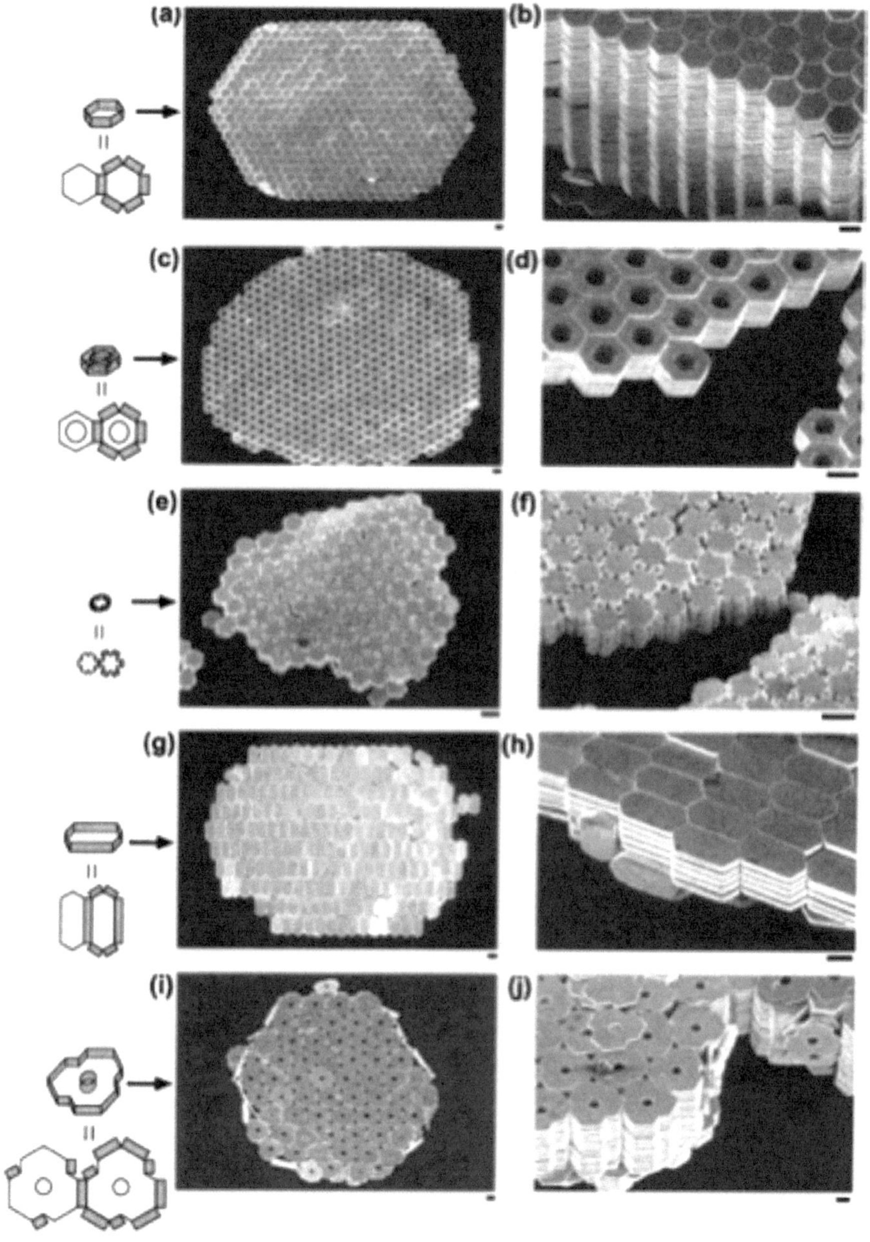

Fig. 10.2 Plates and hexagons made from self-assembled Cr(~OH)|Au(~CH_3)|Cr(~OH). The *scale bar* refers to 10 μm, and *shading* refers to the location of adhesive on each piece. Reprinted with permission from [8]

9]. Working with these larger systems also offers better control of the components and over the interactions among them.

Many types of interactions and mechanisms participate from self-assembly and this concept is used in many disciplines, adapted to their peculiarities.

10.1.1 Structure Formation: Thermodynamics and Kinetics

Structure formation within a system depends on driving forces and on the available paths for minimization of the system Gibbs energy. As a general rule, structures may be formed whenever chemical reactions and mass transfer phenomena are coupled or interdependent. This was first shown by Turing [1] and had important implications, later. For instance, it explains why seeing two or more bands in some electrophoresis, sedimentation or chromatography experiments does not always mean that the experimenter has two or more components in his sample: expected single band for a pure protein may divide in two (or more) bands, depending on the occurrence of association reactions involving the protein molecules [10].

Mass transfer limitations may trap a system in non-equilibrium states, in other ways. One example relevant for the present discussion is the difference between the outcomes of diffusion or reaction-limited aggregation (DLA or RLA) phenomena [11], shown in Figs. 10.3 and 10.4.

Another interesting case is that polymers precipitated from solution often show the initial formation of crystals that are not the most stable ones, but are kinetically favored. The stable crystalline form eventually substitutes for the initial crystals, creating tension within the polymer solid [12].

Self-assembly in the nanoscale is the outcome of the formation of contacts between molecules, particles and ions and this may be represented by changes in surface and interfacial tensions, whenever the involved contact areas are significant. A simple classical example is Adam's [13] description of a macroscopic experiment whose implications for small-scale systems are easily perceived: a block of wax is molten on top of hot water that is cooled back to room temperature. Then the block is removed from the container and the experimenter verifies that its upper surface, which solidified in contact with air is not wetted by water, while water spreads on the lower surface. The minimization of Gibbs energy in the molten block was achieved by accumulating hydrophobic molecules at the upper surface, while hydrophilic carboxylic acid and fatty alcohol accumulated in the interface with water.

An important practical example of the effectiveness of surface tension to direct self-assembly in materials processing is the fabrication of float glass [14]. In the industrial process, molten glass is poured and solidifies over a pool of molten metal, forming flat surfaces that dispense with polishing techniques that were previously used to make glass windows. It is highly probable, although it has not been explicitly demonstrated in the literature, that molten glass and metal acquire opposite charges in the process, that contribute to the formation of a smooth, flat surface.

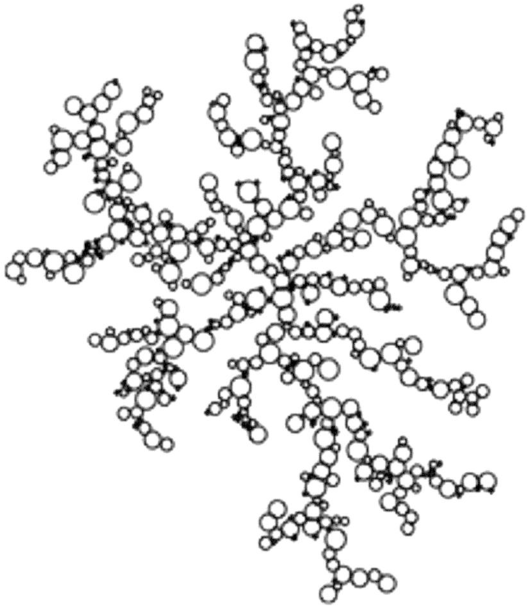

Fig. 10.3 Typical DLA with 104 particles and the size of particles varying from 1 to 5. Reprinted with permission from [11]

10.2 Range and Specificity of Electrostatic Forces

Self-assembly depends on all intervening forces when molecules, ions and particles change place within a system. It differs from the other intermolecular forces by acting at longer distances, producing attraction as well as repulsion. For this reason, patterns of fixed or matching charges, positive and negative, may account for extremely strong and specific interactions. One case that has been studied in great detail is the trypsin–soyabean inhibitor complex, whose association constant is among the higher known for any chemical reactions. The complex has an important contribution from salt bridges formed by charged groups on both proteins, shown in Fig. 10.5 [15].

Moreover, electrostatic interactions have an effect on association kinetics because they can produce *electrostatic steering* that increases association kinetic constants, reaching the limits of the diffusion-limited collision rates [16].

10.3 Biological Systems

The three fundamental noncovalent bonds in biomacromolecules are electrostatic interactions, hydrogen bonds, and van der Waals interactions [17]. The importance of electrostatic forces in biological systems is now well established and this was strongly stated by Politzer et al. [18] as follows: "Noncovalent interactions are predominantly electrostatic in nature. It follows that an effective tool for their

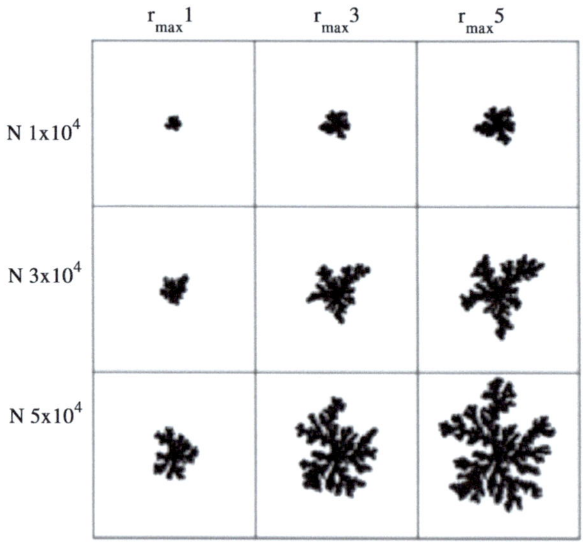

Fig. 10.4 Part of the morphological diagram for different configurations of total number of particles N and maximum radius of particles r_{max} varying from 1 to 5. Reprinted with permission from [11]

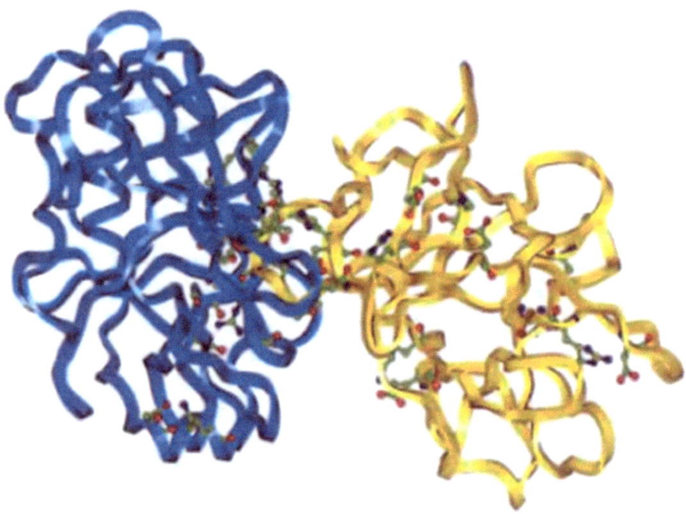

Fig. 10.5 Salt-bridges and inter-molecular hydrogen-bonds in trypsin–soyabean inhibitor complex. Salt-bridges and H-bonds are shown with their side-chains, colored by atom type. Trypsin is to the left. Reprinted with permission from [15]

investigation and elucidation is the electrostatic potential on the molecular surface." These authors developed an approach to obtain a variety of condensed phase macroscopic properties, expressed in terms of site-specific and global statistical quantities that characterize the surface potential. This was initially done on studies on dioxins, anticonvulsants and tetracyclines, the nucleotide bases and reverse transcriptase inhibitors.

Chertsvy [19] recently reviewed many DNA systems where electrostatic interactions play a dominating role. On the other hand, protein engineers can now use the principles of electrostatics and computational protein modeling to stabilize protein interfacial residues and to develop new proteins, targeting biomedical and biotechnological applications [20].

This is now a huge topic and ongoing work goes to great depth that will eventually be used in synthetic nano-sized structures.

10.4 Ionic Surfactants and Polyelectrolytes

"Micelles form to minimize unfavorable hydrocarbon-water interactions" [21]. However, micellization of ionic surfactants forces their headgroups into close proximity creating repulsion. This leads the surface structure of micelles into a compromise: less than half of the hydrocarbon surface in sodium dodecylsulfate (SDS) micelles is covered by polar headgroups leaving room for considerable contact between hydrocarbon and water.

Thus, coulombic interactions contribute to create micellar structures that are more complex than simple schematic pictures, showing the inner hydrocarbon chains shielded by the head groups.

Other effects of electrostatic interactions are observed in surfactant self-assembly. For instance, the system formed by sulfobetaine/pentanol/toluene/water and poly(diallyl-dimethylammonium chloride) (PDADMAC) shows a complex phase diagram, where a transition between phases was explained by a change of curvature of the sulfobetaine surfactant film due to Coulombic interactions with the polycation [22].

Self-assembled polyelectrolyte systems involving lyotropic liquid crystalline and hydrophobic polyelectrolytes, i.e. block polyelectrolytes, associating polyelectrolytes and polysoaps were reviewed [23]. In these cases, self-assembly is created by interactions between the polyelectrolyte and the surfactant molecules, in dilute or semi-dilute solutions, in gels or the solid state, with the participation of both attractive and repulsive Coulombic interactions.

10.5 Colloidal Crystals, Macrocrystals and Co-crystals

Many authors have described the formation of ordered arrays of uniformly sized polymer or silica particles. Poly [styrene-co-(2-hydroxyethyl methacrylate)] forms core and shell latex particles that easily undergo self-arraying forming *colloidal crystals* or *macrocrystals*, evidenced by the iridescence of dispersions and dry solids that is in turn due to Bragg diffraction of light ordered particles, whose sizes are in the same range as the wavelength of light [24]. Different crystallographic arrangements coexist in the dry latex and fracture surfaces (Fig. 10.6) [25].

The ease of particle self-assembly in the latex dispersions and dry solids obtained therefrom is assigned to two factors: (a) the existence of repulsive interactions among particles together with attractive interactions between the particles and the liquid, driving particle ordering at long distances within the dispersion, while particle diffusion throughout the dispersion is still possible; (b) the hydrophilic surface

Fig. 10.6 Secondary electron scanning micrographs of PS/HEMA latex film showing fracture surfaces. HCP and FCC crystalline domains are marked in micrograph **c**. The *bars* represent the scale in micrometers. Reprinted with permission from [25]

layer, which allows for strong capillary adhesion during the latex drying process, according to Nagayama's model [26].

Electrostatic self-assembly drives co-crystal formation of apoferritin protein cages with poly(amidoamine) dendrimers (PAMAM), with very large crystal domain sizes. PAMAM dendrimers with generations from two to seven were used to produce the co-crystals, whose crystal structure and lattice constant showed a dependency on dendrimer generation. Co-crystals currently show a great potential for making protein-based mesoporous materials, nanoscale multicompartments and metamaterials [27].

10.6 Micro- and Nano-fabrication Through Electrostatic Self-assembly

Surface patterning based on electrostatic adhesion has been used with great success and many beautiful examples come from the Whitesides group. Patterned surfaces were successfully used to assemble 10-μm diameter gold disks on planar and curved substrates. Surface charge patterns were first generated either by microcontact printing or by photolithography. Gold disks fabricated by electroplating into photoresist

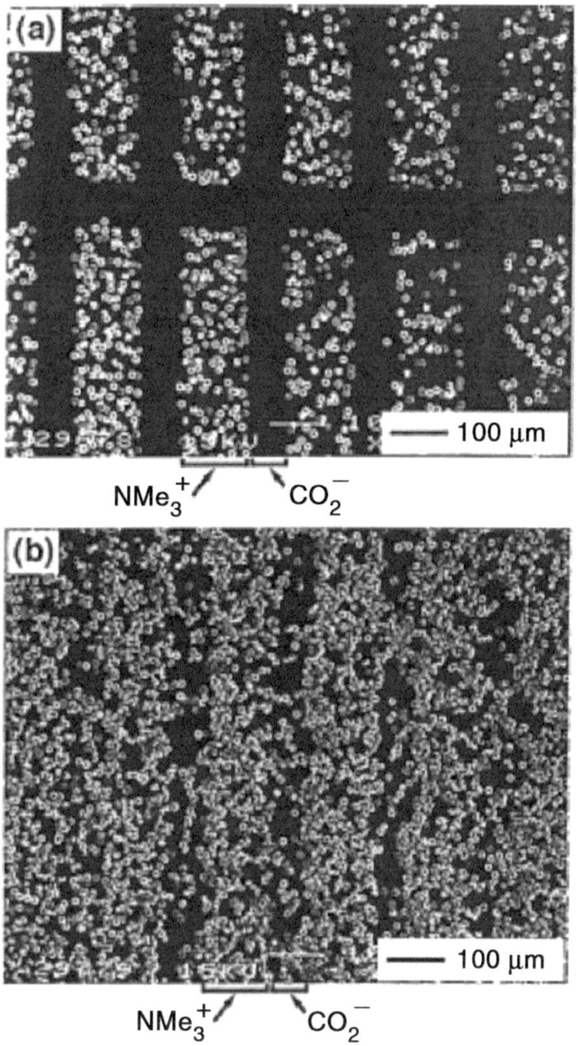

Fig. 10.7 SEMs of HPO_3^--terminated gold disks, assembled onto patterned gold wafers. (**a**) Assembly in ethanol was selective. (**b**) Assembly in water was not selective. Reprinted with permission from [28]

molds were coated with charged self-assembled monolayers. Adding disk suspensions to the patterned surfaces led to the deposition of aggregates of disks with positive charge on phosphonate-, carboxylate-, and SiOH-terminated surfaces but not on trimethylammonium-and dimethylammonium-terminated surfaces. Negative disks showed the symmetric behavior, and methyl- and trifluoromethyl-terminated surfaces resisted deposition of disks of either charge. Disk deposition was selective and dense in methanol, ethanol, 2-propanol and dioxane but it was nonspecific in water (Fig. 10.7) [28].

Nanocomposite structured materials made from structured gold nanoparticle/polyhedral oligomeric silsesquioxane (POSS) showed an interesting

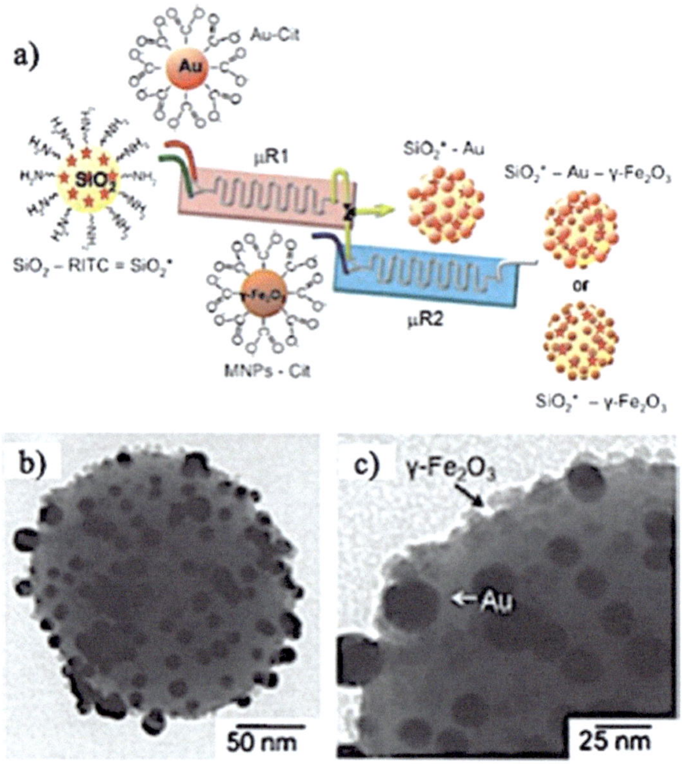

Fig. 10.8 (**a**) Scheme of a two-step microfluidic synthetic procedure for the assembly of multi-functional nanoparticles/fluorescent silica sphere assemblies. *RITC* rhodamine isothiocyanate, *Cit*, citrate. (**b**) and (**c**) TEM images of SiO_2*–Au–γ-Fe_2O_3 nanostructures. Reprinted with permission from [30]

outcome of self-assembly: the increase in spacing within the nanocomposite influenced the surface plasmon resonance band of the larger gold particles. In this work, particles were functionalized with carboxylic and quaternary ammonium groups that provided the opposite charges necessary to form well-organized arrays [29].

10.6.1 Microfabrication Coupled to Microfluidics

Microfluidic techniques allow the control of kinetic aspects of the self-assembly of molecular amphiphiles by the adjustment of the hydrodynamics of the fluids. Several reports show one-step direct self-assembly of different building blocks with a range of products without templates, showing the possibilities of microfluidics for application in various areas like micromotor fabrication and controlled drug delivery [30] (Fig. 10.8).

A review [31] of methods that have been developed for improving microscale assembly, the basic assembly theory and modeling methods proposes three basic

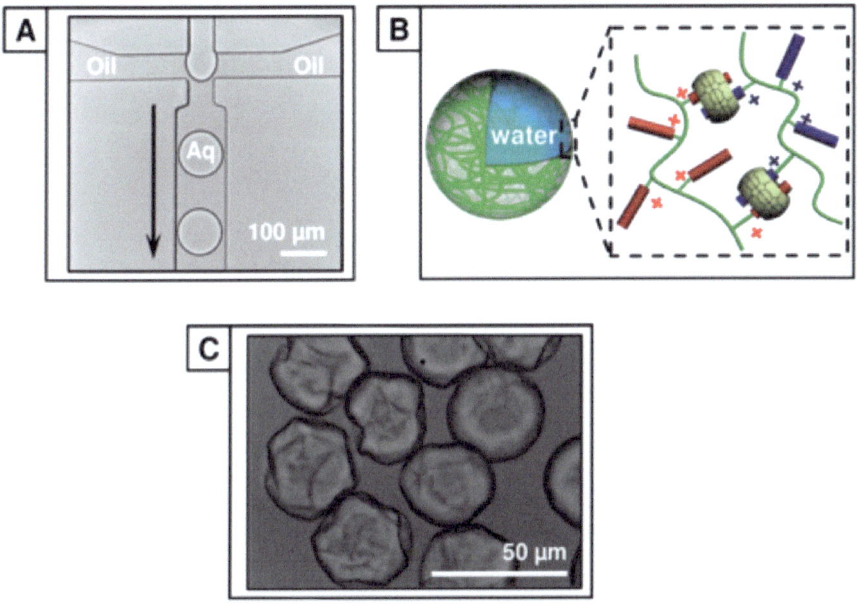

Fig. 10.9 Formation of microcapsules from microfluidic droplets. (**a**) Transmission optical image of the generation of monodisperse water-in-oil microdroplets at a microfluidic flow-focusing junction. (**b**) A schematic of supramolecular microcapsule formation between charged copolymers at the interface of a microdroplet. (**c**) Transmission image of partially collapsed, ultrathin polymer microcapsules. Reprinted with permission from [32]

assembly strategies (tool-directed, process-directed, and part-directed) for categorizing these methods. It reports progress in using fluidic forces including surface tension and external fields (magnetic, electric, light) to aid microscale assembly.

A versatile dynamic assembly methodology was recently described, giving birth to a new generation of self-assembled nanostructures [32]. Capsule-forming components are directed to accumulate to the droplet interface, using electrostatic interactions. Charged copolymers are thus selectively partitioned with the help of a complementary charged surfactant, followed by supramolecular cross-linking done using cucurbit[8]uril. This process is employed to selectively form hollow, ultrathin microcapsules and solid microparticles, from a single solution. The ability to dictate the distribution of a mixture of charged copolymers within the microdroplet is the basis of the single-step fabrication of core-shell microcapsules (Fig. 10.9).

10.7 Conclusions

Electrostatic forces make decisive contributions to the self-assembly of elaborate functional structures in biological systems and in the nanostructured materials that have been created in many laboratories throughout the world, in the past two decades. Given the breadth of the occurrence of self-assembly in natural phenomena, it can be speculated that electrostatic phenomena may also take place in self-assembly within large-scale systems in the planetary scale, as for instance the oceanic currents.

References

1. Turing AM (1952) The chemical basis of morphogenesis. Philos Trans R Soc Lond Ser B 237:37–72
2. Prigogine I, Nicollis G (1977) Self-organization in nonequilibrium systems: from dissipative structures to order through fluctuations. Wiley, New York
3. Liesegang RE (1896) Naturwiss. Wochenschr 11:353
4. Ohkawa M, Yamashita Y et al (2000) Hematite in pyrophyllite ore deposits, Shobara district, southwestern Japan. Miner Petrol 70:15–23
5. Wolfram S (2002) A new kind of science. Wolfram Research, Champaign
6. Thurmer DJ, Bufon CC et al (2010) Nanomembrane-based mesoscopic superconducting hybrid junctions. Nano Lett 10:3704–3709
7. Whitesides GM, Grzybowski B (2002) Self-assembly at all scales. Science 295:2418–2421
8. Clark TD, Tien J et al (2001) Self-assembly of 10-μm-sized objects into ordered three-dimensional arrays. J Am Chem Soc 123:7677–7692
9. Whitesides GM, Boncheva M (2002) Beyond molecules: self-assembly of mesoscopic and macroscopic components. Proc Natl Acad Sci U S A 99(8):4769–4774
10. Cann JR (1970) Interacting macromolecules: the theory and practice of their electrophoresis, ultracentrifugation, and chromatography. Academic, New York
11. Braga F, Mattos OA et al (2015) Diffusion limited aggregation of particles with different sizes: fractal dimension change by anisotropic growth. Phys A 429:38–34
12. Esperidião MCA, Galembeck F (1993) Polypropylene (high density polyethylene) precipitation from stirred solutions. Eur Polym J 29(7):993–997
13. Adam NK (1941) The physics and chemistry of surfaces, 3rd edn. OUP, Oxford
14. Whitesides GM (1995) Self-assembling materials. Sci Am 273(3):146–149
15. Mitra RC (2010) Understanding the role of electrostatics on protein-protein binding. M.Sc. thesis, Clemson University
16. Kastritis PL, Bonvin AMJJ (2013) On the binding affinity of macromolecular interactions: daring to ask why proteins interact. J R Soc Interface 10:20120835
17. Berg MJ, Tymoczko JL et al (2011) Biochemistry. Palgrave
18. Politzer P, Murray JS et al (2001) Molecular surface electrostatic potentials in relation to noncovalent interactions in biological systems. Int J Quantum Chem 85:676–684
19. Cherstvy AG (2015) Electrostatic interactions in dense DNA phases and protein-DNA complexes. In: Schulz N (ed) Advances in electrostatics. Clanrye, Jersey, chapter 1
20. Khan IJ, Stapleton et al. (2015) Electrostatics in protein engineering and design. In: Schulz N (ed) Advances in electrostatics. Clanrye, Jersey, chapter 3
21. Stokes RJ, Evans DF (1997) Fundamentals of interfacial engineering. Wiley, p 217
22. Koetz J, Günther C et al (2003) Polyelectrolyte-induced structural changes in the isotropic phase of the sulfobetaine/pentanol/toluene/water system. Prog Colloid Polym Sci 122:27–36
23. Kötz J, Kosmella S et al (2001) Self-assembled polyelectrolyte systems. Prog Polym Sci 26:1199–1232
24. Teixeira Neto E, Leite CAP et al (2000) Latex fractionation by sedimentation and colloidal crystallization: the case of poly(styrene-co-acrylamide). J Colloid Interface Sci 231:182–189
25. Cardoso ALH, Leite CAP et al (1998) Easy polymer latex self-assembly and colloidal crystal formation: the case of poly[styrene-co-(2-hydroxyethyl methacrylate)]. Colloids Surf A Physicochem Eng Asp 144:207–217

26. Denkov ND, Velev OD et al (1993) Two-dimensional crystallization. Nature 361(6407):26–26
27. Liljeström V, Seitsonen J et al (2015) Electrostatic self-assembly of soft matter nanoparticle cocrystals with tunable lattice parameters. ACS Nano 9(11):11278–11285
28. Tien J, Terfort A (1997) Microfabrication through electrostatic self-assembly. Langmuir 13:5349–5355
29. Carroll JB, Frankamp BL et al (2004) Electrostatic self-assembly of structured gold nanoparticle/polyhedral oligomeric silsesquioxane (POSS) nanocomposites. J Mater Chem 14:690–694
30. Wang L, Sánchez S (2015) Self-assembly via microfluidics. Lab Chip 15:4383–4386
31. Crane NB, Onen O et al (2013) Fluidic assembly at the microscale: progress and prospects. Microfluid Nanofluid 14:383–419
32. Parker RM et al (2015) Electrostatically directed self-sssembly of ultrathin supramolecular polymer microcapsules. Adv Funct Mater 25:4091–4100

Chapter 11
Tribogenerators

Contents

11.1 Introduction

Electrostatically driven motors and generators have been described since the seventeenth century, preceding the electromagnetic equipment that is part of current life. Electrostatic devices participate from many important technologies; they are attractive for hobbyists and ideal for introductory science classes. Even simple equipment produces voltages in excess of 1000 V, with impressive effects. On the other hand, the current and power obtained from electrostatic machines are much lower than those supplied by electromagnetic power supplies.

An early electrostatic storage device was the Leyden jar invented in eighteenth century, a capacitor used in some experiments in electricity that reached some practical use. The early development of the electrical telegraphy [1] was in great part done using static electricity stored in Leyden jars to send messages, until they were replaced by electrochemical devices and later by electricity from the grid. Faraday worked intensely to understand the mechanisms whereby charge was accumulated in Leyden jars. He concluded that it was stored in the glass and not in the water, as believed before. Leyden jar was an essential part of the setup built by Hertz to verify Maxwell's theory of electromagnetism. In 1886, Hertz noticed that discharging a

F. Galembeck, T.A.L. Burgo, *Chemical Electrostatics*,
DOI 10.1007/978-3-319-52374-3_11

Leyden jar through one coil of a Riess spiral produced a spark in the other coil [2]. This was the basis of the *spark-gap transmitter* used by Hertz to prove Maxwell's concepts.

Another device that had practical use was the *Wimshurst machine*, an electrostatic generator based on a rotary disk. A Wimshurst machine is shown in Fig. 11.1. It is one of the so-called *influence machines* that produce electricity by induction, different from the electrostatic friction machines. It was the first electrostatic device that could produce continuous current at the voltages needed for the development of commercial X-ray [3] generators. The latter were in turn simple and robust tools for producing radiation and they were used by many important scientists and Nobel Laureates as Rutherford and Lenard. Nowadays, technical applications that require high voltages rely on electronic and electromagnetic equipment.

When electrostatic machines were becoming popular in Europe, William Thomson (later Lord Kelvin) was already working intensely in his electrostatic experiments. Among many other important things, he invented methods to measure static electricity still widely used today, as the *Kelvin vibrating capacitive sensor*. He measured atmospheric electrical parameters, contributing to a long-standing active research area in geosciences. He idealized a very simple but robust apparatus to investigate charges in the atmosphere, the *water-dropper equaliser* [4] and the famous *Kelvin water dropper* [5], also referred to as the "Kelvin thunderstorm". He could then prove the existence of charge in the atmosphere even during fair weather, discovering the permanent electric field gradient at the Earth surface, as high as 100 V/m. The Kelvin water dropper itself is a type of electrostatic generator and its miniaturization into a microfluidic device [6] was proven suitable for certain micro applications.

Fig. 11.1. Wimshurst machine used in a teaching laboratory in the Physics Department in the Federal University of Santa Maria

11.1.1 Particle Accelerators: Van De Graaff and Pelletron Generators

Amazing scientific discoveries were made at the beginning of the twentieth century and the structure of atoms was then established, based on particle physics. Charged particles were accelerated by large electric fields and they were made to collide with other particles or targets, affording powerful tools for nuclear science and other fundamental problems in the study of matter. Although great progress was made with the electrostatic machines available at that time, they were far from reaching the ever-increasing desired voltages and this was a limiting factor for progress in high-energy physics, requiring the development of more powerful particle accelerators.

It was well known that the mechanical transmission belts used in conveyor systems could produce electrostatic discharge, building undesirable high voltages. This is the reason why nowadays these belts contain anti-static materials as carbon black and they move on grounded conductive rollers [7]. Having this in mind, in 1931 the physicist Van de Graaff, then working at Princeton University, developed an electrostatic generator to supply suitable potentials for nuclear research with a simple and inexpensive design [8]. It was built from a moving endless belt made with silk, rubber or electrical insulation paper, accumulating electric charge on spherical conductive terminals supported by insulating columns [9]. The Van de Graaff generators could easily reach 5 million volts across the isolated terminals [9] and they were successfully employed for research on resonance disintegrations and multiply charged ions [10]. They also were used for many applications as in food preservation and the sterilization of various biological materials [11, 12], treatment of water and sewage [13]. They are still valuable today for scientific experiments [14] and electrostatic demonstrations.

An improvement over the Van de Graaff type generator was made using a chain rather than regular belts. The newer Pelletron generators were made of chains of metal pellets, joined by insulating nylon links, that are charged by induction [15] dispensing with the rubbing contacts or corona discharges used in Van de Graaff generators. Pelletron generators reach 25 million volts, producing highly uniform ion beams [16] and they have been used for producing secondary beams of unstable nuclei [17], to investigate silicon–germanium heterojunction bipolar transistors (SiGe HBTs) [18] and other applications of ion beam interactions with materials and atoms.

The Van de Graaff and Pelletron generators were the last in a long series of historical and successful electrostatic machines that were superseded by electromagnetic generators as power sources.

11.2 Energy Harvesting and Scavenging

The relationships between magnetism and electricity discovered following the observations made by Hans Christian Orsted and others since 1820 [19] and the theory of electromagnetism by James Clerk Maxwell provided powerful methods to convert mechanical energy to electricity. Electromagnetic generators operate at high power efficiency rates, they produce stable voltage, large current and commercially significant amounts of electric power that the electrostatic generators never matched.

On the other hand, electrochemical batteries became the main devices for electricity storage. This now has a fast-growing demand from portable electronic ware and electric autos, which is being largely answered by lithium batteries. However, battery pricing and availability drives the search for new methods to produce and store energy. For this reason, *energy-harvesting* techniques are actively investigated to provide alternatives to batteries in electronic devices [20].

Energy-harvesting devices show promising features to enable the operation of low-energy consumption devices. Smart skins [21], wireless sensors with zero−/low-power operation and the implementation of the Internet-of-Things (IoT) and of machine-to-machine (M2M) communication [22] could largely benefit from self-powered systems operating continuously for as long as the materials last.

Electromagnetic generators powered by mechanically wasted energy have been created in order to harness energy available in the environment. Vibration energy harvesting has been considered in many applications for wireless and self-powered microsystems [23]. For instance, El-Hami et al. developed a device capable of generating electrical power from mechanical energy in vibrating environments [24]. Schematically shown in Fig. 11.2, the design utilizes an electromagnetic transducer and its operating principle is based on the relative movement of a magnet pole with respect to a coil; it is capable of producing more than 1 mW within a volume of 240 mm^3 at a vibration frequency of 320 Hz. Beebly et al. produced an electromagnetic generator that operates from a vibration level of 60 mg, delivering 46 μW to a resistive load of 4 kΩ when the device is shaken at its resonant frequency of 52 Hz [25]: 30% of the power supplied by the environment is converted to electrical power in the load. Moreover, even the human motion is being considered as a power source for microgenerators, where energy is captured into the system by the inertia of a moving mass within the generator frame [26]. In this case, mechanical motion is used to generate electrical energy through an electric or magnetic field, or a piezoelectric

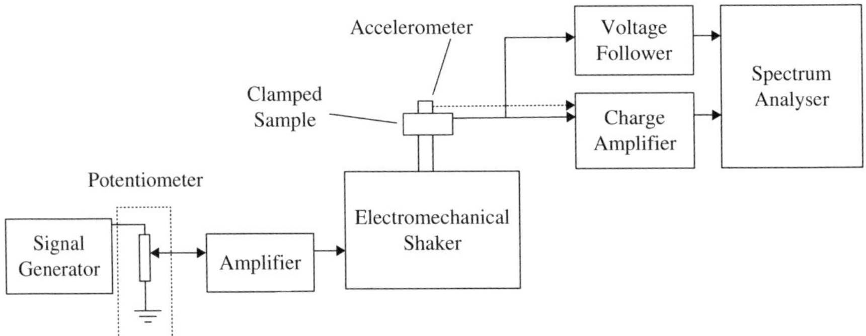

Fig. 11.2. Experimental setup for generating electrical power from mechanical energy in a vibrating environment. Reprinted with permission from Ref. 24

material [27]. An early demonstration of the concept were wristwatches that collected mechanical energy from the user's motion.

Piezoelectric materials have been pointed out as potential candidates for energy-harvesting devices. There is rich information on the development and modeling of piezoelectric transduction for vibration-to-electric energy conversion. Roundy and Wright developed a device (1 cm^3 in size) capable of generating a power output of 375 μW from a vibration source of 2.5 m s^{-2} at 120 Hz. This is sufficient to power a custom-designed 1.9 GHz radio transmitter [28]. Piezoelectric devices have also been tested in the fabrication of a shoe capable of generating and accumulating energy dissipated during walking [29, 30]. Although piezoelectric devices can provide an alternative power source to operate certain types of sensors/actuators, telemetry and MEMS devices, the energy produced is far too small to directly power most portable electrical devices or to recharge batteries [31].

Thermal energy is another energy source for microgenerators. A thermoelectric microconverter was fabricated using thin films of bismuth and antimony tellurides [32]. Using temperatures ranging from 27 to 36 °C found in different parts of a body, a simple thermoelectric microconverter step-up circuit with 1 cm^2, provides a power output of 18 mW, successfully applied to individual electroencephalogram (EEG) modules composed by an electrode, processing electronics, and an antenna [32]. A microgenerator composed of n-type and p-type Bi_2Te_3 nanowire array thermoelectric material was designed to power miniaturized microelectromechanical systems, MEMS [33]. Seebeck coefficients for the p-type and n-type Bi_2Te_3 nanowire arrays are about 260 and −188 μV/K separately with the nanowire diameter ca. 50 nm at 307 K, making these attractive energy sources for MEMS and lab-on-a-chip devices [33].

Finally, some groups have also been proposing multifrequency energy converters where power is generated by conversion of radio-frequency (RF) waves to electrical energy. A modified form of existing CMOS-based voltage doubler circuit achieved 160% increase in output power over traditional circuits [34]. Another promising method for RF energy converting was proposed by Mi et al. [35], who used multiple energy-harvesting antennas in the same area to increase the power/area ratio: increasing area by 83% increased power output by 300%. The energy harvested by RF converters is in the range of a few μW/cm^2 [36]. On the other hand, although the power output generated by RF energy harvesting techniques is small, it is enough to drive low power consumption devices, independent of the grid.

11.3 A New Age for Electrostatic Generators

The development of electrostatic generators was a dormant research area until the 2000s, analogous to many other topics in electrostatics. Then, many scientists and engineers focused their attention on electrostatic phenomena bringing new concepts and design [37] to circumvent the problem of low power output. This brought rapid progress to this field that soon reached a new level of maturity.

Major developments are in small-scale energy harvesting, based on *piezoelectric* and *electrostatic transducers* that are well suited for microelectromechanical system (MEMS) implementation [20].

Micropower generator devices based on electrostatic [38] or electromagnetic [39] forces to convert mechanical energy into electricity have relative advantages and disadvantages but at small scales electrostatic forces tend to be superior to electromagnetic forces [40]. A major difference is that electrostatic forces have significant effects not only on conductors but also on dielectric materials.

Usually, electrostatic generators used to convert pressure variations to energy provide higher density of power than the electromagnetic devices [41]. Many types of systems integrated in footwear, in the floor, attached to human body or set on an environment have been described as capacitor or battery chargers. Using a variable capacitor, Miyazaki et al. fabricated an electrostatic generator with a power output of 120 nW and conversion efficiency of 21% [42]. This is a self-sustained power generator that could scavenge the energy of vibration such as found in small buildings. This vibration-to-electric energy is achieved with a variable capacitor formed by a plate and a fixed electrode. The plate is mounted on a wall that resonates due external vibrational forces, resulting in a periodic change in the capacitance. Another variable-capacitance-type generator was proposed to supply electrical power to a heart pacemaker, its design is shown in Fig. 11.3 [43]. This device generated a power output of ca. 58 μW, supplied to a 1 MΩ load.

There are many other cases showing the use of variable capacitors to convert mechanical energy to electrical energy through changes in capacitance. These devices can be useful for low-power autonomous systems but they are far from reaching power density sufficient for standard applications.

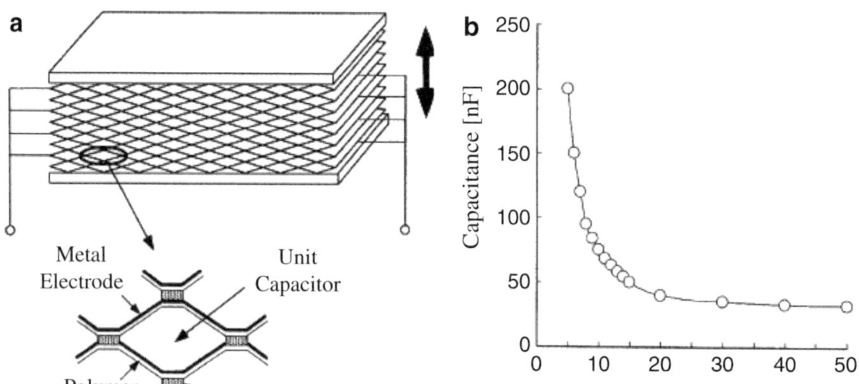

Fig. 11.3. (**a**) Fundamental structure of the honeycomb-type variable capacitor and (**b**) the relation between length and capacitance. Reprinted with permission from Ref. [43]

11.3.1 Nanostructured Electrostatic Generators

Progress in micropatterning, self-assembly and other small-scale techniques created many possibilities to produce a wide range of assembled materials, both conductors and dielectrics, with smooth or very rough surfaces. A leading group is headed by Professor Wang, from the School of Materials Science and Engineering of Georgia Tech. For instance, piezoelectric zinc oxide nanowire (NW) arrays successfully converted nanoscale mechanical energy into electrical energy [44]. Nanogenerator devices were developed using vertically aligned ZnO arrays, delivering 58 V open circuit voltage and 12 μA/cm^2 short circuit current density [45].

Triboelectricity easily affords potentials in the thousand-volts range, higher than piezoelectricity. One of the first triboelectric prototype devices developed in Wang's group reached 230 V output voltage, 15.5 μA/cm^{2}current density, and 128 mW/cm^{3}energy volume density [46]. To achieve such figures the authors used nanostructured polydimethylsiloxane (PDMS) surface with pyramidal patterns and aluminum foil surface with cubic patterns. As seen in Fig. 11.4a, the nanotribogenerator is based on an arch-shaped assembly where PDMS and Al are fixed at the opposite

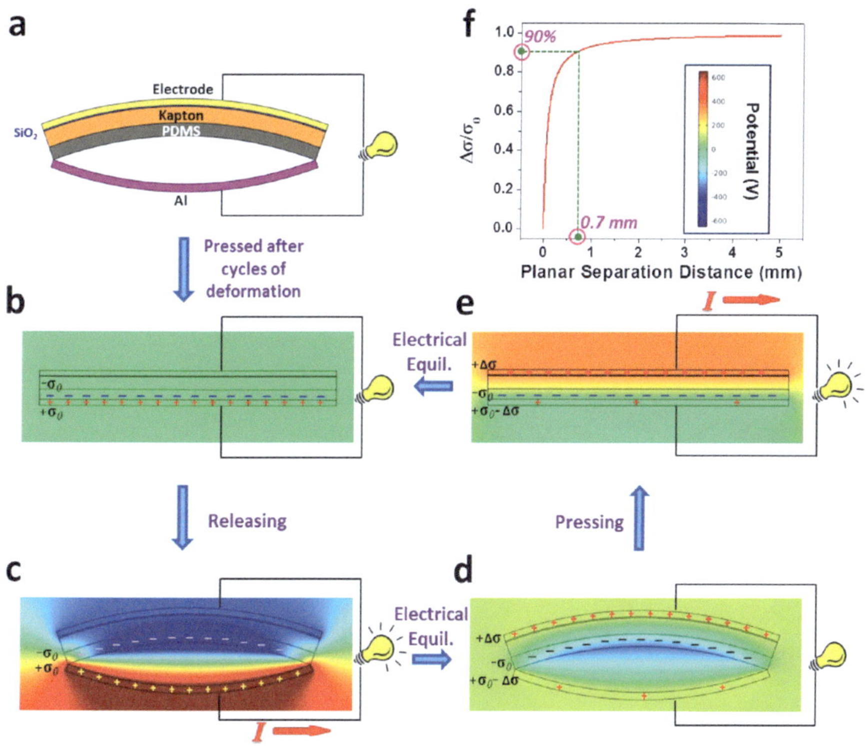

Fig. 11.4. Working principle of the polymer–metal-based tribogenerators. Reprinted with permission from Ref. [46] that describes the detailed authors' explanations

sides. When the device is pressed (Fig. 11.4b–d), there is contact electrification of the two surfaces and charge is transferred to the Al electrode. Thus, mechanical energy is converted to electrical energy.

Using the same principles, Wang's group also described a flexible triboelectric generator with patterned surfaces but now using only polymers for contact electrification. In this case, polymer films made of Kapton polyimide and polyester (polyethylene terephthalate-PET) [47], as shown in Fig. 11.5. Deformation of this device drives electrons toward the external load by changes in capacitance of electrodes fixed at the back of the polymers (details of the charging mechanism will be addressed in the next section). Contact electrification results in a flow of electrical current with an output voltage of up to 3.3 V at a power density of ~10.4 mW/cm^3.

This work was followed by extensive publication describing many different devices: transparent triboelectric generators [48]; generators triggered by human

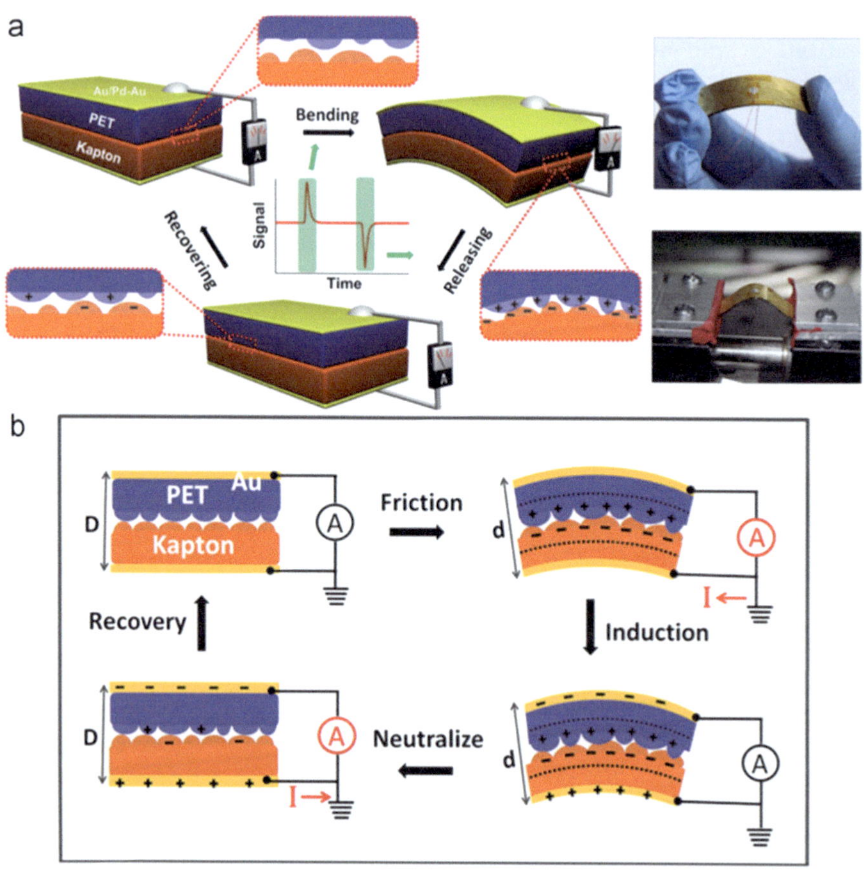

Fig. 11.5. Schematic illustration of the structure and working principle of Wang's triboelectric generator. Reprinted with permission from Ref. [47] that describes the detailed authors' explanations

footfall [49] or based on sliding electrification [50], for harvesting wind energy [51] and a hybrid piezo/triboelectric generator [52]. Progress in device performance is outstanding, regarding both output power surface density and energy conversion efficiency. A thin-film micro-grating triboelectric generator produced 50 mW/cm^2 with conversion efficiency of 50%; other devices reached 85% energy conversion efficiency, when operated at low frequency [53]. Many other groups developed new types of tribogenerators, often based on the setups developed by Wang's group, contributing to the large number of recent publications [54–60].

11.4 Some Fundamental Aspects of Tribogenerators

Most tribogenerators operate according to the principles of influence machines, which means that electrostatic charges generated by friction, induce ("*influence*") charge in a conductor positioned somewhere close to the electrostatic source, like the Wimshurst and Van De Graaff generators.

The new *nanotribogenerators* share some features with the classical influence machines. They produce charges by friction (or contact) and the electrostatic potential is used to induce charge in electrodes. They only differ from traditional electrostatic influence machines in the way they transfer charge to the power output [61].

In the earlier triboelectric generators, surface roughness plays an important role, inspiring nanopattern designs that enhance the real contact area and consequently the effectiveness of triboelectrification, as in the patterned polydimethylsiloxane (PDMS) used in many tribogenerators [46, 56, 58].

The operation of nanotribogenerators requires effective coupling of contact charging and electrostatic induction: for this reason, contacting materials are chosen from opposite ends of the triboelectric series and mounted on separated metal sheets or plates connected to a capacitor.

There is still much room for further progress in the development of nanotribogenerators, because many authors have not yet benefited from the recent information showing nonhomogeneous triboelectric patterns in polymer surfaces. It is now clear that the electrostatic patterns developed after rubbing or contacting dielectric surfaces are never homogeneous, both at macro- [62] and nanoscales [63]. Tribocharges are far from being uniformly distributed in most dielectric surfaces, different from what is as often shown in schematic drawings, positive on one surface and negative on the other. Even though a net charge is observed in tribocharged surfaces, this is an algebraic sum of coexisting positive and negative charges and the overall charge density is thus less than density in small areas. Consequently, induced charge in the collecting conductors is also lower than what could be obtained by achieving an effective spatial separation of positive and negative charges. This is perhaps a next step in improving nanotribogenerators: controlling and perhaps using charge mosaic formation in the contacting surfaces, maximizing charge separation and thus energy conversion efficiency.

However, at this point in time this is more easily said than done, because in most systems that were studied in detail the mosaics formed are chaotic-deterministic [62, 63]. Even so, nanotribogenerators are now a bright reality and research in this area will probably yield many new exciting results followed by their practical applications.

References

1. Huurdeman AA (2003) The worldwide history of telecommunications. Wiley, Hoboken
2. Aitken HGJ (1985) Syntony and spark: the origins of radio. Princeton University Press, Princeton
3. Porter TC (1896) The X-rays produced by a Wimshurst machine. Nature 55:30–32
4. Aplin KL, Harrison RG (2013) Lord Kelvin's atmospheric electricity measurements. Hist Geo Space Sci 4(4):83–95
5. Thomson W (1867) On a self-acting apparatus for multiplying and maintaining electric charges, with applications to illustrate the voltaic theory. Proc R Soc Lond 16:391–396
6. Marín AG, van Hoeve W, García-Sánchez P, Shui L, Xie Y, Fontelos MA, Eijkel JCT, van den Berg A, Lohse D (2013) The microfluidic Kelvin water dropper. Lab Chip 13:4503–4506
7. Hart SL, Daenzer KP et al (2008) Low electrostatic discharge conveyor. US Patent 7,328,785 B2, 12 Feb 2008
8. Van De Graaff RJ (1931) A 1,5000,000 volt electrostatic generator. Phys Rev 38:1–2
9. Van Atta LC, Nothrup DL, Van Atta MC, Van De Graaff RJ (1936) The design, operation, and performance of the Round Hill electrostatic generator. Phys Rev 49:761–776
10. Wells WH (1938) Production of high energy particles: a review of the recent progress in the development of extremely high voltages. J Appl Phys 9:677–689
11. Wertheim JH (1960) Radiation sterilization of fluid food products. US Patent 2,962,380 A, 29 Nov 1960
12. Trump JG, Van De Graaff RJ (1948) Irradiation of biological materials by high energy roentgen rays and cathode rays. J Appl Phys 19:599–604
13. Dunn CG (1953) Treatment of water and sewage by ionizing radiations. Sewage Ind Waste 25(11):1277–1281
14. Ortega-Jimenez VM, Dudley R (2013) Spiderweb deformation induced by electrostatically charged insects. Sci Rep 3:2108
15. Herb RG (1974) Pelletron accelerators for very high voltage. Nucl Instrum Methods 122:267–276
16. Aguiar VAP, Added N et al (2014) Experimental setup for single event effects at the São Paulo 8UD pelletron accelerator. Nucl Instrum Methods Phys Res B 332:397–400
17. Lichtenthäler R, Lépine-Szily A et al (2005) Radioactive ion beams in Brazil (RIBRAS). Eur Phys J A 25:733–736
18. Praveen KC, Pushpa N et al (2012) Application of a Pelletron accelerator to study total dose radiation effects on 50 GHz SiGe HBTs. Nucl Instrum Methods Phys Res Sect B 273:43–46
19. Ørsted HC, Jeveld K et al (1998) Selected scientific works of Hans Christian Ørsted. Princenton University Press, p 421–445.
20. Mitcheson PD, Yeatman EM et al (2008) Energy harvesting from human and machine motion for wireless electronic devices. Proc IEEE 96(9):1457–1486
21. Le T, Lin Z et al (2015) Smart skins: could they be the ultimate sensing tool? IEEE Nanotechnology Magazine, June 2015, 4–10.
22. Tentzeris MM, Georgiadis A et al (2014) Energy harvesting and sensing. Proc IEEE 102(11):1644–1648

23. Beeby SP, Tudor MJ et al (2006) Energy harvesting vibration sources for microsystems applications. Meas Sci Technol 17:R175–R195
24. El-hami M, Glynne-Jones P et al (2001) Design and fabrication of a new vibration-based electromechanical power generator. Sensor Actuators A 92:335–342
25. Beeby SP, Torah RN et al (2007) A micro electromagnetic generator for vibration energy harvesting. J Micromech Microeng 17:1257–1265
26. von Buren T, Lukowicz P et al (2005) Kinetic energy powered computing—an experimental feasibility study. Proc 7th IEEE Int Symp Wearable Comput 22–24.
27. von Buren MPD, Green TC, Yeatman EM, Holmes AS, Troster G (2006) Optimization of inertial micropower generators for human walking motion. IEEE Sensors J 6(1):28–36
28. Roundy S, Wright PK (2004) A piezoelectric vibration based generator for wireless electronics. Smart Mater Struct 13:1131–1142
29. Shenck NS, Paradiso JA (2001) Energy scavenging with shoe-mounted piezoelectrics. IEEE Micro 21:30–42
30. Rocha JG, Goncalves LM et al (2010) Energy harvesting from piezoelectric materials fully integrated in footwear. IEEE Trans Ind Electron 57(3):813–819
31. Sodano HA, Inman DJ et al (2005) Comparison of piezoelectric energy harvesting devices for recharging batteries. J Intell Mater Syst Struct 16:799–807
32. Carmo JP, Gonçalves LM et al (2010) Thermoelectric microconverter for energy harvesting systems. IEEE Trans Ind Electron 57(3):861–867
33. Wang W, Jia F et al (2005) A new type of low power thermoelectric micro-generator fabricated by nanowire array thermoelectric material. Microelectron Eng 77:223–229
34. Jabbar H, Song YS et al (2010) RF energy harvesting system and circuits for charging of mobile devices. IEEE Trans Consum Electron 56(1):247–253
35. Mi MH, Mickle MH et al (2005) RF energy harvesting with multiple antennas in the same space. IEEE Antennas Propag Mag 47(5):100–106
36. Hagerty J, Helmbrecht F et al (2004) Recycling ambient microwave energy with broad-band rectenna arrays. IEEE T Microw Theory 52(3):1014–1024
37. Paramo GL (1991) Rolling triboelectric generator. US Patent 4,990,813, 5 Feb 1991
38. Mitcheson PD, Sterken T, He C, Kiziroglou M, Yeatman EM, Puers R (2008) Electrostatic microgenerators. Meas Control 41:114–119
39. Arnold D (2007) Review of microscale magnetic power generation. IEEE Trans Magn 43(11):3940–3951
40. Mitcheson PD, Green TC (2012) Maximum effectiveness of electrostatic energy harvesters when coupled to interface circuits. IEEE Trans Circuits Syst I Regul Pap 59:3098–3111
41. Harb A (2011) Energy harvesting: state-of-the-art. Renew Energy 36:2641–2654
42. Miyazaki M, Tanaka H et al (2003) Electric-energy generation using variable-capacitive resonator for power-free LSI: efficiency analysis and fundamental experiment. Proc Int Symp Low Power Electron Des:193–198
43. Tashiro R, Kabei N et al (2000) Development of an electrostatic generator that harnesses the motion of a living body. JSME Int J Ser C 43(4):916–922
44. Wang ZL, Song J (2006) Piezoelectric nanogenerators based on zinc oxide nanowire arrays. Science 312:242–246
45. Wang ZL, Zhu G et al (2012) Progress in nanogenerators for portable electronics. Mater Today 15:532–543
46. Wang S, Lin L et al (2012) Nanoscale triboelectric-effect enabled energy conversion for sustainable powering of portable electronics. Nano Lett 12:6339–6346
47. Fan FR, Tian ZQ et al (2012) Flexible triboelectric generator. Nano Energy 1:328–334
48. Fan FR, Lin L et al (2012) Transparent triboelectric nanogenerators and self-powered pressure sensors based on micropatterned plastic films. Nano Lett 12:3109–3114
49. Zhu G, Lin Z et al (2013) Toward large-scale energy harvesting by a nanoparticle-enhanced triboelectric nanogenerator. Nano Lett 13:847–853

50. Zhu G, Chen J et al (2013) Linear-grating triboelectric generator based on sliding electrification. Nano Lett 13:2282–2289
51. Yang Y, Zhu G et al (2013) Triboelectric nanogenerator for harvesting wind energy and as self-powered wind vector sensor system. ACS Nano 7:9461–9468
52. Jung WS, Kang MG et al (2015) High output piezo/triboelectric hybrid generator. Sci Rep 5:9309
53. Wang ZL, Chen J, Lin L (2015) Progress in triboelectric nanogenerators as a new energy technology and self-powered sensors. Energ Environ Sci 8:2250–2282
54. Zhang XS, Han MD et al (2013) Frequency-multiplication high-output triboelectric nanogenerator for sustainably powering biomedical Microsystems. Nano Lett 13:1168–1172
55. Meng B, Tang W et al (2014) A high performance triboelectric generator for harvesting low frequency ambient vibration energy. Proc IEEE Int Conf Micro Electro Mech Syst:346–349
56. Ko YH, Nagaraju G et al (2014) PDMS-based triboelectric and transparent nanogenerators with ZnO nanorod arrays. ACS Appl Mater Interfaces 6:6631–6637
57. Guo H, Chen J, Tian L, Leng Q, Xi Y, Hu C (2014) Air flow-induced triboelectric nanogenerator as a self-powered sensor for detecting humidity and air flow rate. ACS Appl Mater Interfaces 6:17184–17189
58. Kim BH, Barnhart BS, Kwon JW (2015) Electrostatic power generation using carbon-activated cotton thread on textile. Micro Nano Syst Lett 3:3
59. Guo H, Leng Q, He X, Wang M, Chen J, Hu C, Xi Y (2015) A triboelectric generator based on checker-like interdigital electrodes with a sandwiched PET thin film for harvesting sliding energy in all directions. Adv Energy Mater 5:1–9
60. Cui N, Liu J, Gu L, Bai S, Chen X, Qin Y (2015) Wearable triboelectric generator for powering the portable electronic devices. ACS Appl Mater Interfaces 7:18225–18230
61. Wang ZL (2013) Triboelectric nanogenerators as new energy technology and self-powered sensors—Principles, problems and perspectives. ACS Nano 7(11):9533–9557
62. Burgo TAL, Ducati TRD, Francisco KR, Clinckspoor KJ, Galembeck F, Galembeck SE (2012) Triboelectricity: macroscopic charge patterns formed by self-arraying ions on polymer surfaces. Langmuir 28:7407–7416
63. Baytekin HT, Patashinski AZ, Branicki M, Baytekin B, Soh S, Grzybowski BA (2011) The mosaic of surface charge in contact electrification. Science 333:308–312

Chapter 12
Accidents and Losses Caused by Electrostatic Discharge

Contents

Fundamental aspects of this subject and a wealth of practical information are presented in many specialized books. Taylor and Secker [1], Luttgens and Wilson [2], Kaiser [3], Jones, King and Yablonski [4] authored books on the industrial problems caused by static electricity. They cover the basic electrostatic theory, describing the initiation of electrostatic phenomena, measurements of field, voltage and charge and finishing with a detailed chapter on electrostatic discharges. A recent book is devoted to the static ignition hazards in chemical operations [5] and further specific references are given in Sect. 12.6.1, on dust explosions.

F. Galembeck, T. A. L. Burgo, *Chemical Electrostatics*,
DOI 10.1007/978-3-319-52374-3_12

This chapter brings complementary information, largely from recent publications. It also benefits from the information on charging mechanisms that was produced during the last 20 years, with an emphasis on the contribution made by ions to charge accumulation on insulators.

12.1 Electrostatic Discharge in Contacting Points

Electrostatic discharge (ESD) refers to the sudden and often unwanted transfer of static charge between objects at different electrostatic potentials. It is largely associated with high voltage but many experimenters have observed that it can take place under low voltages. This was explored in a systematic study on very low voltage electrostatic discharge done using a small storage capacitor of 2.2 pF and a small relay. The authors analyzed contact discharge current waveform and estimated parameters such as contact resistance, line inductance and total capacitance, identifying the contact electrostatic discharge as gas-ionization discharge and tunnel current discharge, experimentally. The corrosion effect of the contact electrodes on the electrostatic discharge current and the optical emission were also examined [6].

This places electrostatic discharge between electrified metal surfaces within a well-established theoretical framework. However, mass transfer may also take place when one of the electrified surfaces is liquid or when it is powdered. This is seen in Figs. 12.1 and 12.2 that show mass transfer between electrodes. In the first case, one of the electrodes is a bed of carbon powder and in the second it is deionized, distilled water.

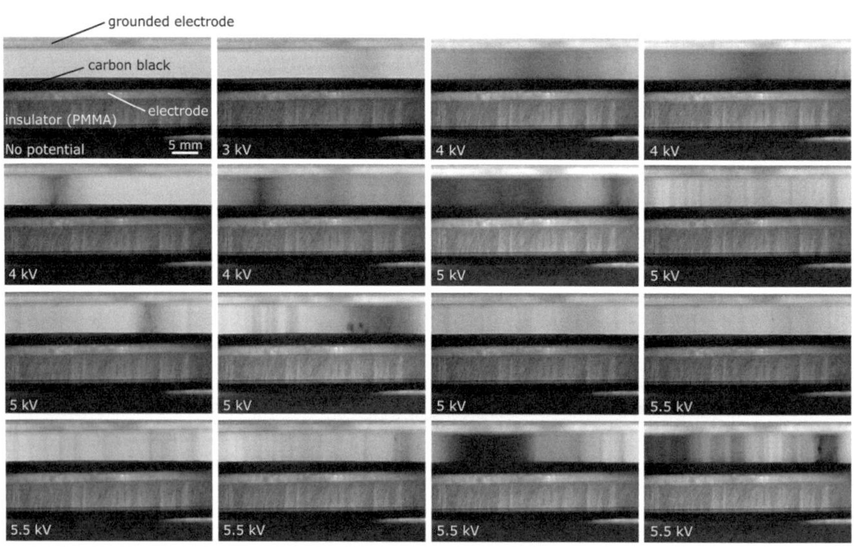

Fig. 12.1. Carbon black mass transfer. When a voltage is applied to the bottom electrode (covered with a layer of carbon black particles), carbon particles move to the upper, grounded electrode. This material transfer is done only through some *channels* that are formed between the electrodes

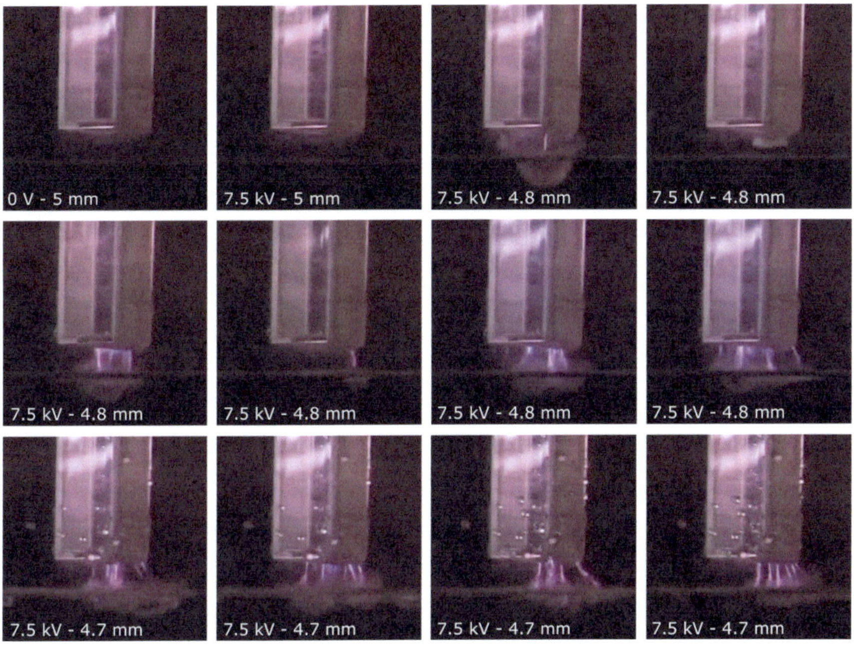

Fig. 12.2. Electrostatic discharge in deionized water. When an electric potential is applied to a steel electrode (supported and molded around a plastic cuvette), electrostatic discharge occurs between the electrode and grounded deionized water, depending on the potential difference and on the distance between the electrode and water

Many accidents, especially "powder explosions" were observed with powdered dielectrics under relatively high humidity, for instance during ship loading operations with flour and grain. Under these conditions, the ejection of particles between electrified bodies cannot be left out of consideration.

12.2 Electrical Discharge in the Electronics Industry

ESD belongs to a family of electrical problems known as electrical overstress (EOS). It poses a serious threat to electronic devices, such as microcircuits, transistors, and diodes that have been examined in great detail [7]. It affects the operation of the systems built using these devices and most electronic companies regard all semiconductor devices as ESD sensitive. For this reason, ESD is a major concern in the microelectronic and electronic industry, in manufacture and testing. ESD is also a source of concern in non-electronic components such as disk drives, magnetic recording heads, and sensors. The emergence of the "Internet of Things" (IOT) placing billions of sensors and devices everywhere only adds to its importance. Its causes, damaging effects and how its effects can be prevented or minimized have been carefully examined [8].

Electrostatic discharge protection in semiconductors underwent both revolutionary and evolutionary changes to maintain pace with the rapidly scaling and changing environment of advanced semiconductor technologies. As a result, it is an area of continuous discovery in both semiconductor development and design and ESD remains a research and development arena for new innovation, invention, and initiatives [9].

ESD protection circuits often use the silicon-controlled rectifier (SCR)-based devices [10] in CMOS ICs but there are also practical problems limiting their use. This requires additional modified device structures and circuit techniques to achieve effective on-chip ESD protection in CMOS ICs [11].

The ESD problem has become a challenging reliability issue in nanometer circuit design. High voltages resulted from ESD might cause excessive current densities in a small device, thus requiring on-chip protection circuits for IC pads. To reduce the design cost, the protection circuit should be added only for the IC pads with an ESD current path, which arise the need of ESD current path analysis. Work has been done for circuit modeling as a constrained graph, decomposition of ESD connected components linked with the pads, and application or search procedures that lead to the identification of the ESD-connected components and thus the current paths. An algorithm for ESD detection path is available to the public for the ESD analysis [12].

The effects of destructive and nondestructive electrostatic discharge (ESD) events applied either to the gate or drain terminal of MOSFETs with ultrathin gate oxide, emulating the occurrence of an ESD event at the input or output IC pins, respectively, were also investigated. ESD stresses may modify the MOSFET current driving capability immediately after stress and during subsequent accelerated stresses. The damage introduced by ESD in MOSFETs increases when the gate oxide thickness is reduced [13].

The ESD sensitivity of 65-nm fully depleted SOI MOSFETs (with thin silicon body) used as output buffer devices was studied in order to classify the observed failure modes and mechanisms leading to the proposal of a new failure criterion [14].

Susceptibility scanning is increasingly adopted for root cause analysis of system-level immunity sensitivities. It allows localizing affected nets and integrated circuits and it can be used to compare the immunity of functionally identical or similar ICs or circuit boards. This methodology and its limitations are discussed in the literature [15].

12.2.1 Role of the Human Body

An experimental study of the ESD characteristics for the charged human body approaching electronic circuit packs was reported. Calculations based on measured capacitance coefficients provide a comparison of the relative probability and severity of the ESD events [16].

12.2.2 ESD and RF Devices

Radio Frequency Identification (RFID) is now widely used for wireless data transmission and reception in applications including automatic identification, asset tracking and security surveillance [17].

An extensive electrical characterization of radio frequency micro electromechanical systems (RF-MEMS) switches has been carried out to identify the dynamic response of devices, driven in different conditions. The authors found that an optimum actuation voltage must be chosen as a tradeoff between good switch transmission and isolation properties and the need to avoid bouncing phenomena when the actuation voltage has been applied. The authors also investigated the reliability of the RF-MEMS switches, finding that it could be heavily affected by ESD phenomena [18].

12.3 ESD in Lighting Equipment

The electrostatic discharge (ESD) properties of the InGaN-light emitting diode (LED) were investigated and the LEDs with higher internal capacitance were found to be more resistant to external ESD impulses [19].

Efforts are described in the literature to obtain LEDs with improved ESD resistance, by changing the GaN layer growth parameters [20].

12.4 Crushing and Milling

Comminution operations often lead to electrostatic charging in the equipment and in the output product, as well. An impressive report shows that electrostatic discharge caused fire and explosion accidents at a sulfur factory in Shanghai. This work describes the test result of potential at the factory workshop, and presents several protective measures in order to prevent dangerous discharge [21].

12.5 Failure of Home, Office and Field Personal Equipment

Many persons have had personal or home equipment damaged with no clear indication on the relevant causes, but the literatures shows reports of systematic work on this topic. In one study, a comparison of the SEM EDX spectra for the metal charge contacts from field-returned cordless phones and for laboratory samples of fixed gap discharge (FGD) tests reveals similar chemical elements, suggesting that ESD is the possible source for cordless phone field failure [22].

12.6 Electrostatic Charge Ignition

All explosives, under all conditions must be considered vulnerable to generation, accumulation and uncontrolled discharge. Energy static hazards as low as 2–3 mJ need to be guarded against. The hazard is normally associated with manufacturing and filling operations due to static charge accumulated on a person supplying energy up to 20 mJ. Electrostatic sensitivity tests [23] can provide an important input regarding electrostatic hazards. A study of electrostatic sensitivity data of some of the initiatory explosives such as nickel/cobalt hydrazinium nitrate, silver azide, lead azide and mercury salt of 5-nitro tetrazole together with data for samples coated with polyvinyl pyrrolidone showed that the electrostatic spark sensitivity of primary explosives decreased in the order of AgN_3 = NiHN > PbN_6 > MNT > CoHN > BN CP. The polyvinyl pyrrolidone-coated samples followed the same order but with increased spark sensitivity [24].

ESD sensitivity of solid propellants and the adaptation of the handling and use procedures have always been an essential part of the safety analysis of rocket systems. However, incidents or accidents have occurred during the past decades, showing that the appreciation of the risks and the knowledge of the mechanisms of ignition are still insufficient. These incidents always involved hydroxyl-terminated polybutadiene-based aluminized composite propellants. Even though much work was done to establish sensitivity criteria, the evaluation of the level of sensitivity of a sensitive propellant has remained as a difficult problem [25].

Ignition of non-aerosolized powders by ESD was investigated using a spherical powder of Mg, for which thermal ignition kinetics was described in the literature. The experimental setup was built based on a commercially available apparatus for ESD ignition sensitivity testing and the authors found that the ignition is primarily due to direct Joule heating of the powder by the spark current [26].

Another study was on titanium powder heating and ignition by an ESD or spark. Different ignition modes were observed for powders with different morphologies and placed in layers with different thickness. For both spherical and sponge Ti powders prepared as monolayers, ESD initiation resulted in fragmentation of the initial particles. Particle fragments were ejected from the sample holder and burned as individual metal droplets. Sponge powder placed in thicker layers ignited generating individual burning particles with combustion times close to those expected based on the particle size distribution. Spherical Ti powder placed in thicker layers was difficult to ignite and only a few short individual particle streaks were observed, corresponding to the finest particles present in the sample. When a titanium powder (either spherical and sponge) was placed in a layer with thickness greater than 0.1 mm, significant fusing of the particles was observed which reduced the powder heating by the discharge's Joule energy [27].

Hydrogen ignition by ESD at a ventilation duct outlet was investigated concerning the effect of the outlet shape. Iron (III) oxide particles were used as the model dust. Grounding the ventilation duct suppresses most of the electrostatic charge, but not all charge generation from the mixture of hydrogen and iron oxide is removed by grounding only [28].

Textiles, leather, or other materials that are used in a pure oxygen environment are in serious risk of being ignited by ESD and may cause fire and even disasters or death of astronauts. The materials are more easily ignited at higher oxygen pressure [29].

Sadly, while this text was being written a serious accident was reported in the United States, inflicting permanent injury to a young student [30].

12.6.1 Dust Explosions

Dust explosions are the main topic of many recent books [31–35] and they are also treated on books on industrial hazards. They are highly feared in grain and flour storage, handling and processing, in the food industry. The first large-scale accident associated with electrostatic discharge was an explosion of flour, on December 14, 1785 in Turin, Italy. Dust explosions are often triggered by ESD and the main relevant parameters are the minimum ignition energy, spark duration time, feeding rate of ignition energy, circuit capacitance, ignition voltage, among others [36].

An extensive list of dust explosions is in a report by Yuan et al. [37] and the cases initiated by sparks or static electricity are abstracted in Table 12.1. A representative report is on the explosion and fire that virtually destroyed the Malden Mills facility in Methuen, Massachusetts, injuring 37 persons (Fig. 12.3) on December 1995 [39]. The plant produced fleece fabrics and the triggering event was likely a dust explosion involving nylon flock fibers, according to reports by OSHA, the Fire Administration (USFA), and Massachusetts State Fire Marshal's Office. There were reports of previous events when nylon fibers were ignited by static electricity, at the same facility, but managers and employees did not generally understand that the fibers were an explosion hazard before the Malden Mills explosion. Other industrial dust explosions investigated by the U.S. Chemical Safety Board (CBS) [38] were: polyethylene dust explosion in Kinnston, NC and aluminum dust explosion in the Huntington, IN (Figs. 12.3 and 12.4).

Table 12.1 Some examples of dust and other explosions reportedly triggered by static electricity [26, 27]

Date	Country	Material	Equipment involved	Types of industries	Ignition source
1960	China	Coal	Electric locomotive	Coal mine	Electrical sparks
1986	Japan	Chemical	Weighing hopper	Chemical plant	Static electricity
1991	China	Coal	Roadway, cable	Coal mine	Electrical sparks
1995	US	Nylon	Textiles machinery	Fleece fabrics	Static electricity
2002	Brazil	Solid fuel	Launching rocket	Satellite	Electrical spark
2007	China	Wood	Packing workshop	Wood products process	Electrical sparks
2008	China	Metal	Dust collection system	Metal fabrication	Electrical sparks
2012	China	Metal	Polishing workshop	Metal polishing plant	Electrical sparks

Fig. 12.3 Picture taken from the site of a factory after an explosion assigned to polyethylene dust. Reprinted with permission from Ref. 38

Fig. 12.4 Fire triggered by an aluminum dust explosion. Reprinted with permission from Ref. 38

12.6.2 Electrostatic Charging Associated with Liquid Leakage

Leakage accidents of pressurized flammable liquids often occur in chemical plants. A study investigated electrostatic hazards due to liquid leakage by observing the amounts of electrostatic charge during the leakage, the electric field of the clouds generated by leakage, various types of pipeline design. Water and kerosene were used as the leaking liquids. The authors found that electrostatic charges depend on leakage parameters such as the type of liquid, the gasket material between flanges, and the pipeline pressure. In all tests, the amount of the electrostatic charges of water was larger than that of kerosene. The maximum value of the electric field, generated from the leakage liquid in this study is a safe level. No incendiary electrostatic sparks, such as brush and/or spark discharges, were detected in these tests [40].

12.7 Protection Against Electrostatic Discharge

Given the importance of ESD, its prevention is often discussed in the literature, in a number of papers [41, 42]. Given its economic and strategic importance in circuitry, greater attention is given to this area.

One paper considers two independent approaches for ESD protection: reducing the likelihood of having an ESD event and improving the robustness of the devices themselves. The first focuses on reducing the amount of charge that is developed and controlling its redistribution. The second approach reviews ways to improve the circuit robustness by improving individual circuit elements and by adding additional elements for charge flow control and voltage clamping [43].

12.7.1 New Materials for Avoiding Electrostatic Discharge

12.7.1.1 Textiles

Clothing and home textiles are frequent sources of electrostatic discharge, since they are usually insulators submitted to mechanical efforts and friction. For this reason, textiles have been developed to make clothing for researchers and engineers handling electrical and electronic devices, and subsystems, but also working in clean-room applications such as the pharmaceutical and optical industries. In general, most man-made fabrics are subject to electrostatic discharge and many approaches were developed to overcome problems caused by such conditions. For instance, stainless steel/polyester woven fabrics made on a handloom were developed successfully. Stainless steel staple fibers are incorporated into these fabrics as conductive fillers to promote the electrostatic discharge properties of the woven fabric. Attenuation of the ESD for various woven fabrics is determined by an ESD immunity tester in a voltage range of 4–8 kV [44].

Poly(vinyl chloride)/graphite nanosheet/nickel (PVC/GN) nanocomposites are other alternative candidates for electrostatic charge dissipation and electromagnetic interference shielding applications due to their lightweight, easy processing and tunable conductivity. The applicability of the nanocomposites in electrostatic charge dissipation was tested in terms of displaying the variation of decay voltage with time and its dielectric properties were investigated. This material also has adequate characteristics to be used in radar evasion [45].

12.7.1.2 Carbon Nanotubes and Graphene

Multiwalled carbon nanotubes (MWNTs), graphene and nanographites are promising alternatives for use in polypropylene (PP) and other thermoplastics and rubber matrices, for electrostatic discharge applications. In some cases, they replace the "carbon blacks" widely used in industry that are not as highly conductive.

In one work, the surfaces of MWNTs were modified with octadecylamine (ODA) via CF_4 plasma-assisted fluorination and subsequent alkylamination. The number of fluorine groups on the MWNT surface was controlled by varying the CF_4 plasma treatment conditions. The resulting MWNT/PP nanocomposites exhibited a much finer dispersion in the insulating PP matrix than was observed for non-modified MWNTs, leading to an enhanced electrical conductance at low MWNT loading (2 wt %). Furthermore, the nanocomposites showed significant increase in the storage modulus (G') and complex viscosity (η^*) in the low-frequency region with MWNT loading, showing a rheological percolation threshold at 1 wt % MWNT content. This method can be applied to the fabrication of other carbon nanotube-based polymer nanocomposites for electrostatic dissipative materials with high mechanical strength and for other high-performance industrial applications [46].

Powder energetic materials are highly sensitive to ESD ignition. The addition of small concentrations of carbon nanotubes (CNT) to the highly reactive mixture of aluminum and copper oxide (Al + CuO) significantly reduces ESD ignition sensitivity. The lowest CNT concentration needed to desensitize ESD ignition is 3.8 vol. % corresponding to percolation producing an electrical conductivity of 0.04 S/cm [47].

The great interest in graphene for nanocircuitry led to comprehensive study of ESD characterization of atomically thin graphene. In a material comprising only a few atomic layers, equivalent current density in the 10^8 A/cm^2 range was observed, for very short times. Failure occurs within the graphene and not at the contacts. Moreover, unique gate biasing effects were observed that can be exploited for novel applications including new ESD protection designs for semiconductor products. Graphene's robustness against high-current/ESD pulses suggests that it has great potential in a variety of applications [18].

12.7.2 Devices and Equipment: Corona Electrodes

Corona electrodes produce clouds of gaseous ions, either positive or negative. These may be deposited on aerosol particles, increasing their collection by electrostatic filters. Corona discharge is thus used as pre-charger in electrostatic precipitators (ESP). The involved phenomena are complex and they are only partially understood, due to overall status of knowledge on static charging. For instance, back ionization of dusts with high resistivity may take place, perhaps contributing to the fouling of discharge electrodes. The literature reports examples of the applications of the direct current corona with spraying discharge electrodes [16].

Precipitator design may broaden its applicability, as in the case of a barrier discharge type ESP which gives a high removal efficiency for both NO_x and particles generated by Diesel combustion [48].

A study with laboratory-scaled wire-to-plate electrostatic precipitator (WPESP). showed that the dynamic air flow velocity modifies the current density and the electric field distributions on the planes surfaces of the WPESP [49]. This result shows that precipitation performance is affected by the flowing gas.

12.8 Safety Codes

Given the widespread impact of ESD, great attention has been paid to reducing risk through best practice. Some relevant documents are listed below.

- Documents on ESD program guidance or requirements: ANSI ESD S 20, 20–2007-Standard for Development of an ESD Control Program, ANSI/ESD S8.1-Symbols-ESD Awareness, or ESD TR20.20-ESD Handbook [50].
- Requirements for specific products or procedures such as packaging requirements and grounding, e.g. ANSI/ESD S6.1-Grounding and ANSI/ESD S541-Packaging Materials for ESD-Sensitive Items [38].
- Standardized test methods used to evaluate products and materials, e.g. ASTM-257-DC Resistance or Conductance of Insulating Materials. The electronics industry originally relied on standards developed for other activities but today specific test method standards are available, largely as a result of the ESD Association's activity. These include standards such as ANSI/ESDA-JEDEC JS-001-2010-Device Testing, Human Body Model and ANSI/ESD STM7.1: Floor Materials-Resistive Characterization of Materials [38].
- ATT-TP-76306 practice on Electrostatic Discharge Control, created within AT&T. The purpose of this practice is to provide an overview of ESD phenomena and how ESD can be controlled in order to improve the performance of AT&T network and to lower operating costs [51].

12.9 Electrostatics and Chaos

A complicating factor in ESD prevention is the possibility to have chaotic-deterministic phenomena taking place under different situations, contributing to the uncertainties that may contribute to electrostatic phenomena. Indeed, there are many reports on oscillation and chaotic behavior in systems subjected to electrostatic forces.

For instance, the formation of complex oscillations was reported for an electrostatic microelectromechanical system: upon increasing an ac voltage, a cascade of period doubling bifurcations took place giving rise to chaotic transition. The nonlinear nature of the electrostatic force was shown to be responsible for the reported observations [52]. Another publication addresses chaotic behavior including its prevention, in micro/nano resonators [53].

The early Kelvin measurements on atmospheric electricity already showed that the atmospheric electric field is a highly variable quantity at different timescales [54] that is suggestive of a fractal behavior. Field mill measurements are currently used to produce time series that contributed to the identification of maximum electric field mill values as a lightning predictor that contributes skill in forecasting the occurrence of lightning [55]. Interestingly, three of four predictors identified are sensitive to moisture.

12.10 Final Comments

ESD prevention is currently based on tested procedures that are often practiced by well-prepared professionals. Nevertheless, accidents reportedly assigned to ESD continue to happen throughout the world, taking a heavy toll in lives and property.

In the authors' view, this situation is due to current status of consolidation of the scientific information on how electrostatic charge builds-up and dissipates, in solids, liquids and gases. The absence of consensus on the involved mechanisms places the accident prevention professionals in the same situation as a medical doctor who has to treat patients of a disease with known symptoms but an undefined etiology, or causation. Moreover, a large amount of careful research was done within a conceptual framework much simpler than what is now known about charge pattern formation in dielectrics. A good example is the large number of references in the admirable Greason's book from 1992, always referring to charge transfer due to unidirectional electron driven by differences in work function [56]. We now know that triboelectricity in insulators is characterized by charge patterns, not by the formation of uniformly charged surfaces. In this situation, all the preventive measures taken may be effective whenever the same causes are operating. However, safety procedures are still largely based on empirical evidence that is not accompanied from powerful models with strong predictive character.

For these reasons, it is essential that recent progress on understanding electrostatic phenomena and the intervening mechanisms continues toward a consolidation of current information, subjecting every piece of information to validation and testing hypotheses, models and theories.

Even more important, the results of this effort need to be disseminated. This book contains many references to high-quality information that was just forgotten or dismissed, few years after the relevant papers were published. Paying due attention to this literature will increase our ability to avoid disasters triggered by electrostatic discharge.

References

1. Taylor DM, Secker PE (1994) Industrial electrostatics (electronic & electrical engineering research studies). In: Hughes JF (ed) Electrostatics and electrostatic application series, vol 13. Research Studies Press (Somerset), New York
2. Luttgens G, Wilson N (1997) Electrostatic hazards. Butterworth-Heinemann, Oxford
3. Kaiser KL (2005) Electrostatic discharge. CRC Press, Boca Raton
4. Jones TB, King JL et al (1991) Powder handling and electrostatics: understanding and preventing hazards. CRC Press, Boca Raton
5. Britton LG (2010) Avoiding static ignition hazards in chemical operations. Wiley, New York
6. Oda T, Myasaka H et al. (2009) The low voltage electrostatic discharge on the contacting point. 2009 IEEE Applications Society Annual Meeting. IEEE Industry Applications Society Annual Meeting, p 299–304.
7. Greason WD (1992) Electrostatic discharge in electronics. Research Studies Press (Taunton), New York.
8. Sadiku MNO, Akujuobi CM (2004) Electrostatic discharge (ESD). IEEE Potentials 22:39–41
9. Voldman SH (2004) A review of electrostatic discharge (ESD) in advanced semiconductor technology. Microb Releases 44:33–46
10. Lou L, Liou JJ (2008) An improved compact model of silicon-controlled rectifier (SCR) for electrostatic discharge (ESD) applications. IEEE T Electron Dev 55(12):3517–3524
11. Ker MD, Hsu KC (2005) Overview of on-chip electrostatic discharge protection design with SCR-based devices in CMOS integrated circuits. IEEE T Device Mat Re 5(2):235–249
12. Liu HY, Lin CW et al. (2006) Current path analysis for electrostatic discharge protection. IEEE/ACM International Conference on Computer-Aided Design, p 510–515. http://cc.ee.ntu.edu.tw/~ywchang/Papers/iccad06-esd.pdf. Accessed 13 Sept 2016
13. Cester A, Gerardin S et al (2006) Electrostatic discharge effects in ultrathin gate oxide MOSFETs. IEEE T Device Mat Re 6(1):87–94
14. Griffoni A, Tazzoli A et al. (2008) Electrostatic discharge effects in fully depleted SOI MOSFETs with ultra-thin gate oxide and different strain-inducing techniques. Electrical Overstress/Electrostatic Discharge Symposium Proceedings, pp 59–66.
15. Muchaidze G, Koo J et al (2008) Susceptibility scanning as a failure analysis tool for system-level electrostatic discharge (ESD) problems. IEEE T Electromagn C 50(2):268–276
16. Xu D, Li J et al (2003) Discharge characteristics and applications for electrostatic precipitation of direct current: Corona with spraying discharge electrodes. J Electrostat 57:217–224
17. Bauer-Reich C, Reich M et al. (2011) The interaction of electrostatic discharge and RFID (2011) advanced radio frequency identification design and applications, chapter 8, pp 155–170. http://www.intechopen.com/books/advanced-radio-frequency-identification-design-and-applications/the-interaction-of-electrostatic-discharge-and-rfid. Accessed 13 Sept 2016

18. Tazzoli A, Peretti V et al (2007) Electrostatic discharge and cycling effects on ohmic and capacitive RF-MEMS switches. IEEE Trans Device Mater Reliab 7(3):429–436
19. Jeon SK, Lee JG et al (2009) The effect of the internal capacitance of InGaN-light emitting diode on the electrostatic discharge properties. Appl Phys Lett 94:131106
20. Lai FI, Hsieh YL et al (2011) Enhancement in the extraction efficiency and resisting electrostatic discharge ability of GaN-based light emitting diode by naturally grown textured surface. Diamond Relat Mater 20:770–773
21. Sun K, Hangyu Z et al (2001) Investigation of electrostatics during sulphur crushing operations. J Electrostat 51:435–439
22. Coyle RJ, Jon MC (2000) Electrostatic discharge failure mechanism for cordless phone charge contacts. J Electrostat 49:215–223
23. Fronabarger JW (1996) The electrostatic discharge sensitivity of HMX in the confined state. International Pyrotechnics Seminars 22nd, pp 509–516
24. Wang AZH, Tsay CH (2001) On a dual-polarity on-chip electrostatic discharge protection structure. IEEE Trans Electron Dev 48(2):978–984
25. Davenas A, Rat R (2002) Sensitivity of solid rocket motors to electrostatic discharge: history and future. J Propul Power 18(4):805–809
26. Beloni E, Dreizin EL (2009) Experimental study of ignition of magnesium powder by electrostatic discharge. Combust Flame 156:1386–1395
27. Beloni E, Dreizin EL (2011) Ignition of titanium powder layers by electrostatic discharge. CombustSci Technol 183:823–845
28. Imamura T, Mogi T et al (2009) Control of the ignition possibility of hydrogen by electrostatic discharge at a ventilation duct outlet. Int J Hydrogen Energy 34:2815–2823
29. Wang QG, Zhang X et al (2009) Research on the possibility of ignition of materials by electrostatic discharge in pure oxygen environment at different pressures. J Electrostat 67:876–879
30. Kemsley J (2016) University of Hawaii lab explosion likely originated in electrostatic discharge. Chem Eng News 94(28):5. http://cen.acs.org/articles/94/i28/University-Hawaii-lab-explosion-likely.html. Accessed 13 Sept 2016
31. Cross J, Farrer D (2012) Dust explosions. Springer, New York
32. Field P (2012) Dust explosions. Elsevier, New York
33. Eckhoff RK (2003) Dust explosions in the process industries. Gulf, Amsterdam
34. Amyotte P (2012) An introduction to dust explosions. Elsevier, Amsterdam
35. Barton K (2002) Dust explosion prevention and protection: a practical guide. Elsevier, Amsterdam
36. Nifuku M, Katoh H (2001) Incendiary characteristics of electrostatic discharge for dust and gas explosion. J Loss Prevent Proc 14:547–551
37. Yuan Z, Khakzad N et al (2015) Dust explosions: a threat to the process industries. Process Saf Environ 98:57–71
38. Amyotte PR (2014) Some myths and realities about dust explosions. Process Saf Environ Prot 92:292–299
39. (2006) Combustible dust hazard study. In: Investigation Report No. 2006-H-1, U.S. Chemical Safety and Hazard Investifation Board, chapter 4, p 25. http://www.csb.gov/assets/1/19/dust_final_report_website_11-17-06.pdf. Accessed 13 Sept 2016
40. Choi KS, Yamaguma M et al (2010) Electrostatic charges during liquid leakage. J Loss Prevent Proc 23:294–299
41. Greason WD (1992) Electrostatic discharge: a charge driven phenomenon. J Electrostat 28:199–218
42. Voldman SH (1999) The state of the art of electrostatic discharge orotection: physics, technology, circuits, design, simulation, and scaling. IEEE J Solid State Circ 34(9):1272–1282
43. Vinson JE, Liou JJ (2000) Electrostatic discharge in semiconductor devices: protection techniques. Proc IEEE 88(12):1878–1900
44. Cheng KB, Ueng TH et al (2001) Electrostatic discharge properties of stainless steel/polyester woven fabrics. Text Res J 71(8):732–738

45. Al-Ghamdi AA, El-Tantawy F et al (2009) Stability of new electrostatic discharge protection and electromagnetic wave shielding effectiveness from poly(vinyl chloride)/graphite/nickel nanoconducting composites. Polym Degrad Stab 94:980–986
46. JII L, Yang SB et al (2009) Carbon nanotubes-polypropylene nanocomposites for electrostatic discharge applications. Macromolecules 42:8328–8334
47. Poper KH, Collins ES et al (2014) Controlling the electrostatic discharge ignition sensitivity of composite energetic materials using carbon nanotube additives. J Electrostat 72:428–432
48. Kuroda Y, Kawada Y et al (2003) Effect of electrode shape on discharge current and performance with barrier discharge type electrostatic precipitator. J Electrostat 57:407–415
49. Said HA, Nouri H et al (2015) Effect of air flow on corona discharge in wire-to-plate electrostatic precipitator. J Electrostat 73:19–25
50. (2010) Fundamentals of electrostatic discharge. Part six—ESD standards. ESD Association, New York. https://www.esda.org/about-esd/esd-fundamentals/part-6-esd-standards/. Accessed 13 Sept 2016
51. ATT-TP-76306 (AT&T) (2009) Electrostatic discharge control. https://ebiznet.sbc.com/sbc-nebs/Documents/ATT-TP-76306.pdf. Accessed 13 Sept 2016.
52. De SK, Aluru NR (2005) Complex oscillations and chaos in electrostatic micro-electromechanical systems under superharmonic excitations. Phys Rev Lett 94:204101
53. Tajaddodianfar F, Yazdi MRH et al (2015) On the chaotic vibrations of electrostatically actuated arch micro/nano resonators: a parametric study. Int J Bifurcation Chaos 25:1550106
54. Aplin KL, Harrison RG (2013.) Lord Kelvin's atmospheric electricity measurements https://arxiv.org/pdf/1305.5347.pdf. Accessed 13 Sept 2016.
55. Mazany RA, Businger S et al (2002) A lightning prediciton index that utilizes GPS integrated precipitable water vapor. Weather Forecast 17:1034–1047
56. Greason WD (1992) Electrostatic discharge in electronics. Research Studies Press (Taunton), New York, pp 56–72.

Chapter 13
Electrostatic Processes and Products

Contents

13.1 Industrial Applications of Electrostatics

Electrostatic phenomena are at the core of well-established technologies, like xerography or electrophotography, electrostatic coating, and electrostatic precipitation. More recently, electrostatics has been contributing different solutions to the serious problem of glass panel cleaning in solar farms and in related equipment used in space research. Solutions based on electrostatics are currently being considered to face current problems related to pollution, energy, and resource availability as well as computer technologies.

F. Galembeck, T.A.L. Burgo, *Chemical Electrostatics*,
DOI 10.1007/978-3-319-52374-3_13

Masuda pointed to the characteristics of electrostatic phenomena that make them more or less suitable for different types of applications [1]. For instance, the magnitude of the Coulombic forces is usually proportional to the surface area of the object carrying charge. This means that the motion of macroscopic objects with low surface-to-volume ratios is usually dominated by their mass but electrostatic forces become increasingly important at higher surface areas, with a threshold at ca. 100 μm that is easily reached in particles, sheets and fibers. This explains the enormous importance of electrostatic interactions in the behavior of colloidal and nanostructured matter.

Another important issue is the energy density of the electric field that is usually orders of magnitude lower than the corresponding values for the magnetic field, suggesting that the electrostatic field can never be used for large-scale energy production. However, the current trends towards distributed energy production and environmental energy scavenging alleviate this limitation, impelling efforts in this direction. An exciting example are the nanotribogenerators discussed in Chap. 11.

Electrostatic devices perform well as acoustic-to-electric transducers since they handle small amounts of power. They are thus used as condenser, electret and microelectromechanical system (MEMS) microphones. Electrets form a widespread class of products that is discussed in Chap. 7.

Electrostatic forces are also used in touch displays and haptic systems [2].

13.2 Imaging Technologies

There are different types of electrostatic imaging, including electrophotography, electrostatic printing (as in laser printers) and ink-jet printing.

13.2.1 Electrophotography or Xerography, Laser Printers

Photocopiers and laser printers are currently widely found in homes, offices and industry. They work based on three phenomena: latent image formation on a photoconductive surface, followed by electrostatic adhesion between the charged domains on this surface and toner particles that are finally transferred to the paper surface [3]. These technologies have had an immense role in communications and information management, for more than five decades and they currently play an important role in the dissemination of cultural goods.

The relevant basic phenomena are highly reproducible but the difficulties in fully understanding them are well represented in the following phrase that can be read in the description of Schein's book [3]: "On mastering this material, the reader will have a working knowledge of the physics of the complete electrophotographic process and a detailed knowledge of what is known and not known about its most important subsystem development."

This is now a mature technology but new developments are appearing continuously. For instance, in a recent patent application [4] the electrostatic image carrier is a photoreceptor laminated on at least one surface layer of hydrogenated amorphous silicon carbide or photoconductive layer. One important parameter is the ratio (C/(Si + C)) of the number of carbon atoms and the sum of number of Si and C atoms, in the 0.50–0.65 range. The defect density of the surface layer is 9×10^{18} to 2.2×10^{19} spins/cm^3. Other patents refer to various aspects of the toner composition [5–7]. In one specific case, the main novelty is the addition to the toner of an emulsion of particles of a low MW polyester, that imparts improved crease fix and lowers the minimum fusing temperature, enabling a reduction in fuser energy and enhanced fuser life. The images produced using the toner particles are claimed to have excellent quality [8]. Other patents address magnetic toners [9], the light scanning apparatus [10]. These references are a few examples of the results of an intense R&D activity: the Derwent database lists more than 10,000 patents or patent applications under the keywords "electrophotography" and "toner" only, out of which the leading companies held the following numbers, as of June 2016: Canon, 1801; Ricoh, 1480; Fuji Xerox, 686, Kao, 425; Konica, 381.

13.2.2 Ink-Jet Printers

Ink-jet printers also use electrostatic charging in different ways, to extract liquid from the container cartridge or to deflect the liquid droplets to the desired spots on the paper surface.

For instance, a recent patent application describes a method of driving a hybrid inkjet printing apparatus by applying an electrostatic voltage to ink contained in a nozzle, and applying an ejection voltage so as to eject the ink [11].

The surge of interest in additive manufacture and 3D printing opened a new, large area for the application of electrostatic devices capable to deliver and deflect liquid jets [12], placing the solutes or dispersed particles in the desired spots where they are required to build an object. Beam deflection may also have a contribution from associated magnetic fields [13].

13.2.3 Electrostatic Screen-Printing

Electrostatic screen-printing is used for producing marking or images on cans, bottles, carton boxes and other packing goods. When an optical image is produced on a grid coated with photoconductive film, the grid potential from its voltage source appears on the illuminated portion, attracting the toner particles. The particles passing through the grid mesh is then transferred to the object to produce a pattern. A recent patent [14] describes an apparatus where an electroconductive screen is positioned at the target object of printing but without contacting it. A rubbing sponge is set to rub powder into screen and a direct current power supply is set to supply a voltage to the target object and powder. The apparatus also contains an autorotation mechanism to rotate the rubbing sponge. The applicants target this apparatus for coating powder on target objects, e.g., foodstuff, claiming that the powder adheres more uniformly to the target. They also claim that since the structure of the electrostatic screen printing apparatus is simplified, the cost of the apparatus is reduced.

13.2.4 Electronic Paper

Electronic paper and other displays that are currently widespread are based on electrostatic phenomena, especially electrophoresis [15], electrowetting [16], electrochromism [17], liquid–crystal alignment under a field and microfluidics [18] where the fluid flow may be driven by pressure but also by electrosmosis. Electronic paper is now widely used in e-readers and their success together with the ever increasing demand for display improvements propels the development of flexible, video and color electronic paper products [19].

This requires new development on many areas: new driving systems [20], new materials and better understanding of interfacial charge and its effect on wetting, rheological fluid and particle electrophoretic properties. An especially interesting development is in-plane electrophoresis [21].

13.3 Electrostatic Coating

Industrial coating processes based on electrostatics have been used for decades and their scope and impact have grown steadily due to the many advantages they bring to manufacturers and final product users as well, especially those interested in "greener" products and processes [22–25].

There are two main processes of application of powder coatings: electrostatic coating and the immersion of the objects to be coated in fluidized beds. Powder coating grew steadily as a finishing process due to its unique capabilities to produce high-quality finishes without using undesirable volatile organic compounds (VOC) that contaminate the finished products. It contributes to high production rates and lower costs. In the year 2003 powder coating already covered 15% of the industrial coating market of North America [26] and it has been growing steadily since this time, throughout the world.

In electrostatic coating, the finely ground particles of pigmented resin are dispensed from an electrostatic gun and sprayed over a grounded metal part or a non-conductive material pre-coated with a conductive primer that allows current discharge to the ground. The sprayed particles adhere to the substrate forming a film and the coated object is then transferred to an oven where it is heated. The resin softens forming a nonporous matrix followed by crosslinking that imparts high chemical, thermal and abrasion resistance to the coating.

The adoption of electrostatic coating processes brings many economic advantages to the users: venting and exhaust needs are reduced, the powder coating does not sag or drip and the work area can be cleaned using vacuum-cleaners. One-pass operations are often sufficient and powder losses are a few percent only. Moreover, powder recycling is often feasible, minimizing waste disposal needs.

For these reasons, powder coating holds a strong share in the appliance market and it is also important in the auto and transportation equipment industry as well as in the manufacture of furniture, architectural, urban and outdoor equipment that require high resistance to weather, including fence repair. It can be done in high-throughput production lines by mounting multiple parts in racks. Its application range grows steadily thanks to technological development that allows its utilization with a growing number of substrates and painting powders but small units for home use are also available.

The versatility of electrostatic coating makes it useful in diverse applications like forming layered materials in developmental laboratories, as an alternative to dip-coating, spray-coating and spreading with doctor blades.

The application of electrostatics in both powder and liquid coating can improve the quality of food, such as its appearance, aroma, taste, and shelf life. Coatings are commonly found in the snack food industry, in confectionery, bakery, meat and cheese processing. The most important factors influencing food coating quality are powder particle size, density, flowability, charge, and resistivity, as well as the surface properties and characteristics of the target. The most important factors during electrostatic liquid coating, also known as electrohydrodynamic coating, include applied voltage and electrical resistivity and viscosity of the liquid. Understanding these factors is essential for the design of optimal coating systems [27].

Dry coating technologies are currently used also for making medicines. Powdered coating materials are directly coated onto solid dosage forms without using any solvent and then heated and cured to form a coat. It has many advantages related to avoidance of the use of solvents, when compared to conventional liquid coating. There are several dry coating technologies, including plasticizer-dry-coating, electrostatic-dry-coating, heat-dry-coating and plasticizer-electrostatic-heat-dry-coating. The fundamental principles and coating processes of these technologies, as well as their advantages, disadvantages and commercialization potentials were reviewed [28].

An interesting investigation was made to investigate if electrostatic adhesion is significant in food coating. Different food powders were coated onto different food samples, both electrostatically and nonelectrostatically and the adhesion was measured, when the authors found that electrostatic coating increased the adhesion of most food powders onto most food targets. The effectiveness of the process showed some dependence on the resistivity or oil content of the target, resistivity of the powder and particle size. Under low relative humidity, electrostatic adhesion lasted for several weeks, long enough to be valuable [29].

Another implementation of electrostatic coating is electrophoretic paint, widely used in the production of transportation equipment. Typically, the shaped auto bodies are immersed in the paint dispersion within large tanks and they are biased to a suitable voltage, attracting the dispersed paint particles. Both cathodic and anodic coating are practiced, with the prevalence of the former due to extensive use in the auto industry.

Electrophoretic coatings have uniform thickness without porosity, the coating process is suitable for complex fabricated objects with cavities, coating speed is high, it is applicable to wide range of substrate materials (metals, ceramics and polymers pre-coated with suitable primers), the process is easily automated and thus it is not labor-intensive, it involves lower fire risks due to the use of water as a vehicle for pigment and resin, coating materials utilization is highly efficient and it is suitable to the production of customized micro and nanostructures such as laminates and functional gradients [30]. The adoption and continuous improvement of electrophoretic coating made a great contribution to the increase in auto body life that took place during many recent decades.

13.4 Electrowetting

The shape change of liquid films and drops upon the application of electrostatic potential is the basis for a number of recent inventions targeting a broad range of products and processes.

A patent assigned to the University of Cincinnati describes an electrowetting light valve, where light transmission is controlled by the relative positioning of two immiscible liquids with different light transmittance properties [31]. Figure 13.1 [32] describes schematically the functioning of this light valve.

Another patent that is widely cited describes an electrowetting apparatus comprising voltage supplies connected to lower and upper electrodes and a distribution plate for applying electrical potentials that induce movement of the electrolytic droplet between hydrophobic layers of upper and lower chambers.Electrolytic droplet motion is useful in biochemical assays, including drug research, DNA diagnostics, clinical diagnostics, and proteomics [33].

Another interesting invention is a zoom lens suitable for a miniature camera. It is characterized by lens control through a voltage-controlled electrowetting device that includes two fluids with different refractive indices and their interfaces [34].

Electrowetting is currently the basis for the novel electrowetting display devices. There is an impressive patenting activity related to information displays and a few examples are shown in Table 13.1.

Many other results target fluidic and microfluidic applications, shown in Table 13.2.

All the documents listed in Tables 13.1 and 13.2 are from the year 2016, showing the intensity and diversity of prospective applications of electrowetting. It is impressive that the patent assignees include some current technological leaders in various industrial sectors.

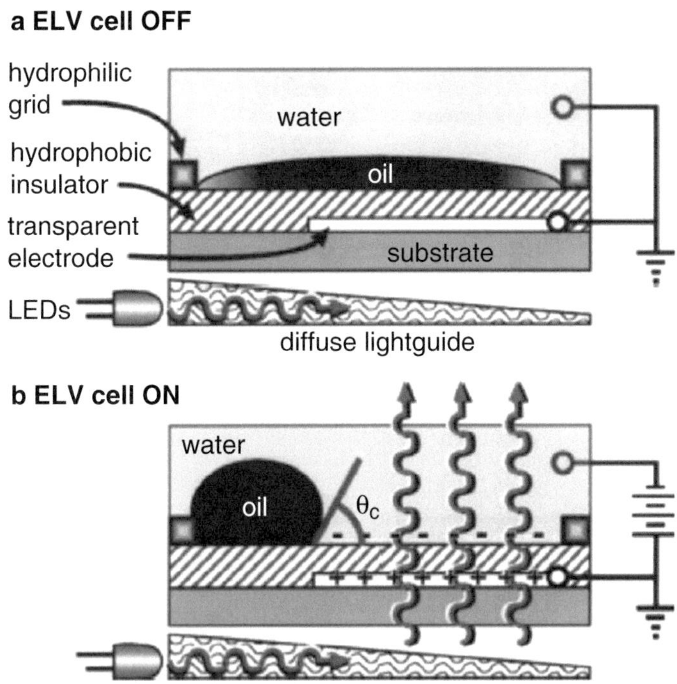

Fig. 13.1 Schematic diagrams of the electrowetting light valve (ELV) device in the OFF (**a**) and ON (**b**) states. Reprinted with permission from ref. [25]

Table 13.1 Some recent patent or patent applications referring to the use of electrowetting in information displays

Patent (granted or application)	Claimed use	Assignee/Inventors
US 9,348,132 B1	Pixel wall and spacer configuration for an electrowetting display	Amazon/Novoselov P. et al.
US2016140692-A1	Method for configuring resolution of a display, i.e., touchscreen display	Google/Pais M. R. et al.
JP2016080728-A	Photosensitive coloring composition for forming colored partition wall of electrowetting device	Mitsubishi/Ito A. et al.
US2016124212-A1 KR2016050397-A	Fluid is used for electrowetting device	SNU R&DB, Samsung and Univ Seoul/Choi C. et al.
US9310602-B1	Color electrowetting display for use in a communication device	Amazon/Chung J. Y.
US2016091711-A1 WO2016050694-A1	Electrowetting element for electrowetting display device used for, e.g., portable, mobile device	Amazon/Tauk L.
US2016088692-A1	Display for an electronic device, includes electrowetting displays	Apple/ Weber D. J. et al.

Table 13.2 Some recent patent or patent applications referring to the use of electrowetting

Patent application	Claimed use	Assignee/Inventors
US 2016/0129437A1	Performing assays in a closed sample preparation and reaction system	Genmark and Adv. Liq. Logic/Kayyem J. F. et al.
US2016151784-A1	Manipulating droplet of liquid or particles in liquid	Univ California/Chiou P. E.
US2016097087-A1	An electrowetting force is applied to move one fluid and a second fluid to a location adjacent the intake tip	Applied Biosystems/ Wiyatno W.
US2016097047-A1	The method is used for growing cells on a droplet actuator. Sample droplet and cell culture medium merge by electrowetting	Advanced Liquid Logic/ Pollack M. G.
JP2016050951-A	Making concave Fresnel lens	Panasonic/Omote A.
KR2016009796-A	Electrowetting anhydrosugar alcohol (anhydrohexitol) in a process step	Samyang Genex/Ryu H.
US2016058272-A1	Posture control of a capsule endoscope by an electrowetting technique	Panasonic/Ishikawa T.
WO2016009114-A1	Electrowetting-based microfluidics-controlled tunable coil to provide a tunable filter or channel selector for a radio	Nokia/Rouvala M. et al.
US2016016170-A1 WO2016011134-A1	Disposable cartridge used in digital microfluidics system for manipulating samples in liquid portion	Tecan et al./Lay T, et al.

13.5 Electrostatic Precipitation

Electrostatic precipitation was the first of the industrial applications of electrostatics and it is now widely used to remove particulate matter from the atmosphere and gaseous effluents in a broad scale range, from household air purifiers to huge systems used, e.g., in thermoelectric power plants [35, 36]. The importance of this technology has been growing for decades, because legislation is continuously reviewed becoming more and more stringent.

Electrostatic precipitation is done by passing the gas stream through a charging device where particles acquire electric charge, followed by a screen or a set of metal bars biased to high voltage, where the particles adhere. Electrostatic precipitators (ESP) can operate with a high collection efficiency and low pressure drop, depending on particle charging and migration velocity of charged particles and it is also applicable to airborne microorganisms [37].

However, their performance deteriorates by abnormal phenomena, including back corona for treating high-resistivity dust and corona quenching for fine dusts that required the development of new technologies. Complex gas-phase electrochemical phenomena also take place in electrostatic precipitators contributing to the simultaneous removal of gaseous pollutants, including dioxin, [38] which is a feared contaminant of gaseous effluents. This shows the possibilities of doing electrochemistry at the solid–gas interface, a topic that is not as familiar as electrochemistry at the solid–liquid interface but has been receiving important inputs [39]. In a study comparing the performance of electrostatic precipitation and filtration, the authors raised the possibility of conversion of mutagenic compounds into less mutagenic forms due to electrostatic effects during electrostatic precipitation [40].

Small-sized electrostatic precipitators were developed for air purification in small volumes including car interiors. A portable, silent, and autonomous electrostatic air sampler targeted the collection of airborne pathogens. As much as 98% of airborne particles from 10 nm to 3 μm are collected [41]. Other reports describe a personal aerosol sampler based on electrostatic precipitation and electrowetting actuation of droplets [42], an integrated microfluidic electrostatic sampler for bioaerosol [43].

13.6 Solar Panel Cleaning

Electrostatic adhesion is often involved with surface soiling in many systems and environments and it is a frequent factor for reducing the transparency of window glass with undesirable effects (Fig. 13.2), including solar energy panel lowered efficiency [44]. This led to the examination of a number of approaches for using electric fields to displace the electrified particles from the surfaces.

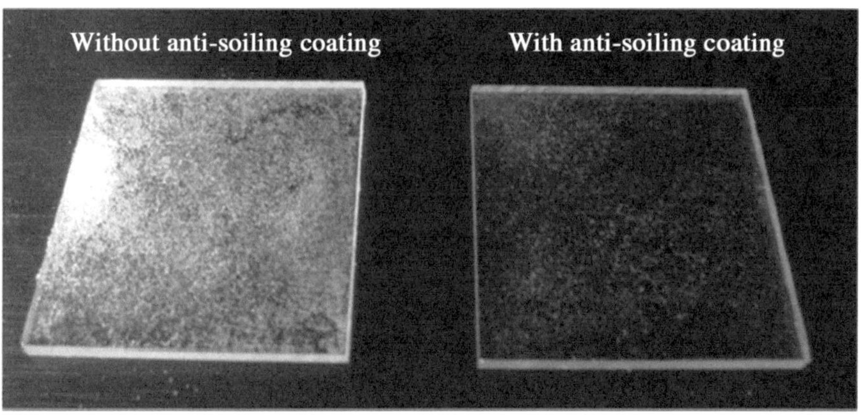

Fig. 13.2 Photographs of two samples with and without an anti-soiling coating, after exposure to impinging sand. Reprinted with permission from Ref. 36

13.7 Electrostatic Separation

13.7.1 Waste Separation

Electrostatic separation has been widely used to separate conductors and nonconductors for recycling e-waste. In a study on the effects of process parameters (voltage used, roller speeds) the separation efficiency of conductors and semiconductors (Fig. 13.3) reached 82.5% and 88%, respectively, showing the feasibility of this technology for recycling valuable metals [45].

13.7.2 Biomass Separation

The recovery of valuable biomass components is often a great challenge in biotechnology. Huge amounts of lignocellulosic materials are obtained as biomass production residues that can be fractionated for preparing fuel. This involves crushing biomass comprising lignin, cellulose and hemicellulose to ultrafine powder, and separating fractions by electrostatic separation of ultrafine powder. This method (Fig. 13.4) enables environmentally friendly dry fractionation of lignocellulosic biomass by economical process, with reduced energy consumption [46].

One case of great current interest is in microalgae production that is intensively examined as a potential source of fuel oil. Microalgae cells are very small, usually

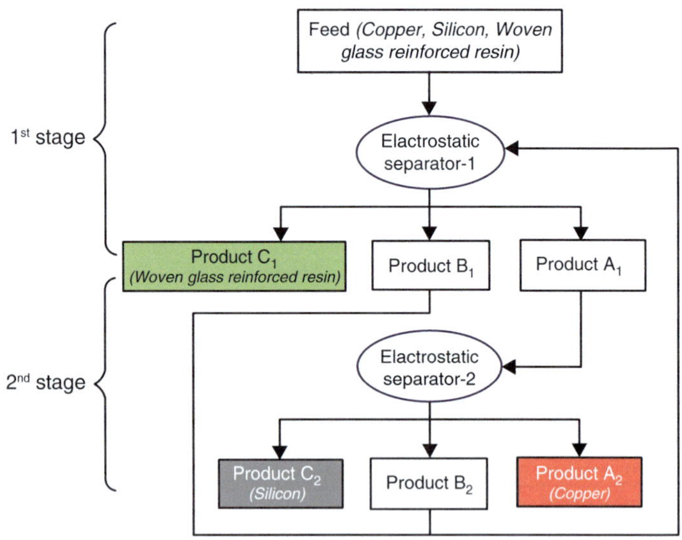

Fig. 13.3 Flowchart of the electrostatic separation of copper, silicon, and woven glass reinforced resin by a two-stage process. Reprinted with permission from Ref. 37

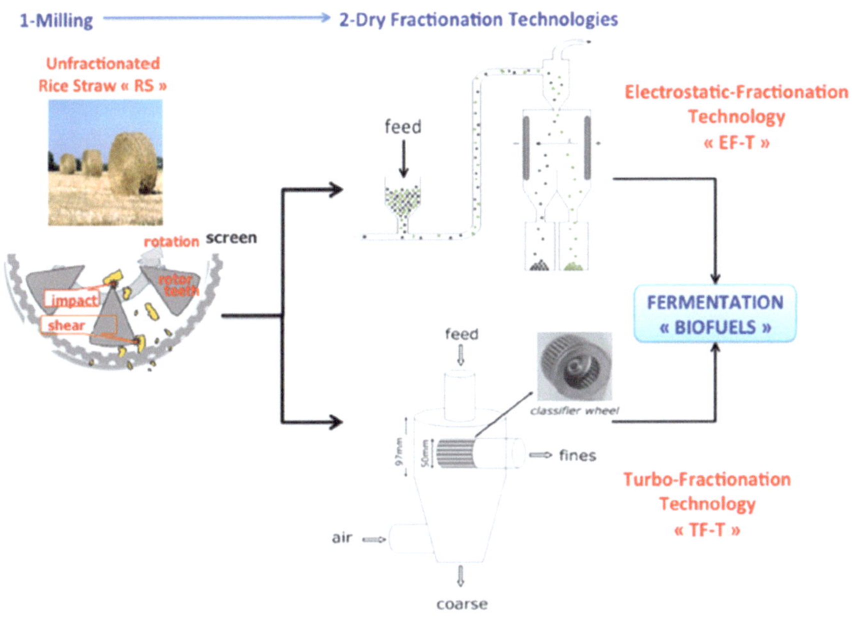

Fig. 13.4 Electrostatic (EF-T) and turbo (TF-T) fractionation technology routes developed for rice straw. Reprinted with permission from ref. 38

less than 10 μm and they grow in highly diluted waters, typically 1 g/L. Microalgae concentration is thus energy intensive and the incurred high costs contribute to make it unsustainable for large-scale production. One particular issue is that large amounts of water require large energy inputs for pumping and dewatering. Microalgae suspension dewatering by electroflocculation requires only 0.3 kWh/m^3 as opposed to 8 kWh/m^3 needed for centrifugation and it does not require large amounts of chemicals, as coagulation, or the large storage volumes that are required for sedimentation under gravity. Achieving optimal results requires careful examination of the various process parameters, including electrode voltage, flow rates and salinity [47].

13.7.3 Electrosorption and Capacitive Deionization

Capacitive deionization (CDI) has attracted the interest of the community investigating water treatment technologies since the mid-1960s. The technology is based on the recognition that high-surface-area electrodes, when electrically charged, can quantitatively adsorb ionic components from water, thereby resulting in desalination. Its theoretical and technological background, the history of its development, and attempts towards scaling up and commercialization have been reviewed, considering its advantages and limitations [48].

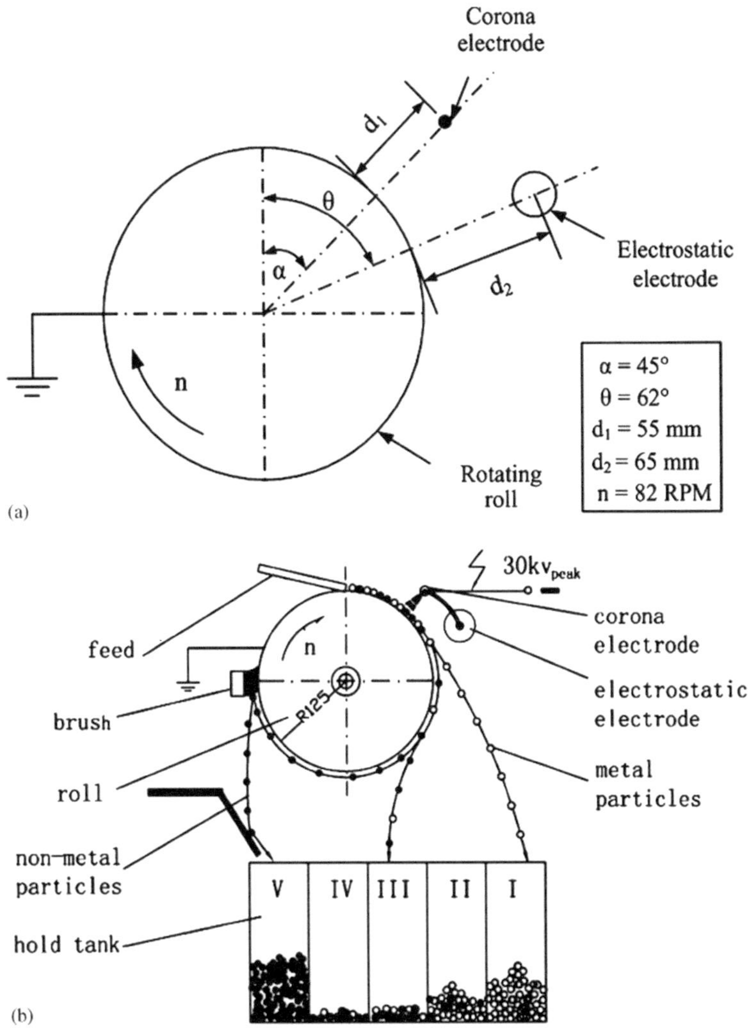

Fig. 13.5 Operating parameters (**a**) of corona electrostatic laboratory separator (**b**). Reprinted with permission from Ref. 41

Some companies have been involved with this technology [40], but it is still a young technology that will probably take some time to reach large-scale application, analogous to what happened to reverse osmosis membranes. Considering the similarity between capacitive deionization systems and supercapacitors, it is not unlikely that a capacitive deionization unit will also have electricity storage capabilities.

13.7.4 Metal Recovery from Electronic Equipment

Printed circuit board (PCB) scrap contains as much as 28% metals, including copper, lead, and tin. Moreover, the content of precious metals in PCBs is more than 10 times that of the exploited minerals. Therefore, the recycling of PCBs is important not only to decrease waste, but also to the recover valuable metals and it is thus an important target of "urban mining." It normally involves mechanical crushing, screening, and drying, to which an electrostatic separation via corona discharge may be added [49]. The results show that the crushing process can completely strip metals from base plates, accompanied by aggregation as opposed to the production on fine powders, concluding that corona electrostatic separation is an efficient and environment-friendly means for recovering metals from PCBs.

13.8 Electroadhesion

Electrostatic grippers are used in the semiconductor industry and in space operations [50] for capture, handling and docking under vacuum. Electroadhesion is applicable to a wide variety of materials and shapes, enabling lightweight, ultra-low power compliant attachment. A typical EA device for space use is made using compliant materials as for instance copper-clad polyimide encapsulated by polymers and arranged as in Fig. 13.6.

Application of electroadhesion covers a broad range of products applicable in advanced manufacture and robotics, as shown in Table 13.3.

The combination of electroadhesion and electrostatic actuation was recently used [51] to handle a wide range of objects, including a fragile water-filled thin

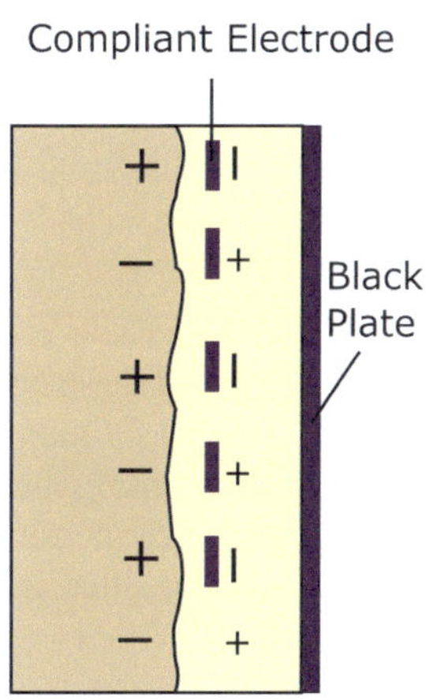

Fig. 13.6 Schematic representation of a compliant electroadhesion clamp. Modified and reprinted with permission from Ref. 50

Table 13.3 United States Patent Trade Office (USPTO) granted patents on electroadhesion

Patent number	Patent title
9,401,668	Materials for electroadhesion and electrolaminates
9,358,590	Electroadhesive surface cleaner
9,308,650	Gripper apparatus
9,302,299	Active electroadhesive cleaning
9,186,709	
9,272,427	Multilayer electrolaminate braking system
9,272,425	Twisted string actuator systems
9,193,402	Structural assessment, maintenance, and repair apparatuses and methods
9,193,068	
9,130,485	Conformable electroadhesive gripping system
9,130,484	Vacuum-augmented electroadhesive device
9,093,926	Electroadhesive conveying surfaces
8,982,531	Additional force-augmented electroadhesion
8,979,034	Sticky boom noncooperative capture device
8,967,548	Direct to facility capture and release
8,861,171	Electroadhesive handling and manipulation
8,833,826	Mobile robotic manipulator system
8,665,578	Electroadhesive devices
8,125,758	
8,564,926	Electroadhesive gripping
8,125,758	
8,325,458	
8,515,510	Electroadhesive medical devices
8,111,500	Wall-crawling robots, wall-crawling devices
7,872,850	
7,554,787	
7,773,363	Electroadhesion
7,551,419	
6,789,679	Method and apparatus for separating particles
6,390,302	
5,839,306	Electronic lock “chiplock”

membrane balloon, a raw chicken egg, a flat sheet of paper, a Teflon tube and a metallic can. All these objects weighted less than 100 g. These impressive capabilities are enabled by maximizing electroadhesion and electrostatic actuation while allowing self-sensing thanks to new dielectric elastomer actuators.

The impressive performance of electroadhesion-based devices and actuators is a strong indication that it will find growing application in advanced manufacture technology.

13.9 Conclusion

Electrostatic phenomena are currently used in many products and processes and their importance and scope is growing, reaching areas as diverse as water and effluent treatment, diagnostics and genomic analysis, printing, coating, robotics and advanced manufacture. There are many exciting new developments like the use of additive manufacture to make electrostatic 3D laser printers and electrostatic motors. However, the applicability of electrostatic processing largely depends on the capabilities of the used materials to interact with electric fields that in turn depends on the materials adequacy to acquire net charge or to undergo sufficiently strong charge induction. It is thus likely that the current developments and better understanding of electrostatic charging, charge dissipation and stability will broaden the scope of electrostatic processing. This will allow the use of materials that do not currently perform well, contributing to a number of new products and related processes.

References

1. Masuda S (1981) Industrial applications of electrostatics. J Electrost 10:1–15
2. Lacroix R, Levesque V (2014) Friction modulation for three dimensional relief in a haptic device. US Patent Application 20140208204, 24 July 2014 (to Immersion Corporation)
3. Schein LB (1992) Electrophotography and development physics. Springer, Berlin
4. Komatsu N et al. (2016) Image formation by electrophotography, involves charging electrostatic image bearing component to charging component, developing latent image with developer including toner and transferring toner image to recording medium. JP2016095471-A, 26 May 2016 (to Canon)
5. Murata S et al. (2016) Manufacture of electrostatic charge image developing toner used in electrophotography, involves mixing aqueous dispersion of wax-containing resin particles containing preset amount of wax, and wax particles, and fusing obtained mixture, JP2016057434-A, 12 April 2016 (to Kao)
6. Maesawa Y (2016) Toner used in image forming apparatus for electrophotography, comprises coloring agent and mixture of amide compound obtained by amidating rosin with amidation agent, and ester compound which is rosin-polyhydric alcohol condensate, JP2016051025-A, 11 April 2016 (to Sharp)
7. Grinwald Y et al. (2016) Liquid toner for printing conductive traces in, e.g., sheet-fed printer, has toner particles dispersed in carrier liquid, where low symmetry electrically conducting material of toner particles is dispersed in resin, WO2016048343-A1, 31 March 2016 (to Hewlett-Packard Indigo)
8. Zhou K et al. (2016) Manufacture of toner particles used for developing images, involves admixing polymeric resin emulsion with crystalline aromatic monoester emulsion including naphthyl benzoate, and adding aggregating agent to obtained composite emulsion, US Patent 9285694 B2, 15 March 2016 (to Xerox)
9. Nishikawa K et al. (2016) Magnetic toner used for electrophotography, comprises magnetic toner particles, binder resin and magnetic material containing inorganic fine particles containing metal oxide, and organic-inorganic composite material fine particles, DE102015112927-A1, 11 February 2016 (to Canon)

10. Park C (2016) Light scanning apparatus for image forming apparatus, has flow restricting unit having opening or transparent element to allow the light beam emitted by light source module to be incident on light deflector and cover covers housing, US Patent 2016070196-A1, 10 March 2016 (to Samsung)
11. Kim JH et al. (2013) Methods of driving hybrid inkjet printing apparatus. US Patent Application 20130176354, 11 July 11 2013 (to Samsung)
12. Chen G et al. (2014) High-voltage static electricity driving and variable diameter 3D printer, has fixing ring formed with cylindrical spring hole one, nozzle end connected with outer screw thread, and needle head connected with high-pressure static generator, CN203611471-U, 28 May 2014 (to the Univ. Beijing Chem. Technology)
13. Byun DY et al. (2015) Electrostatic force patterning apparatus for three-dimensional patterning apparatus used for ink-jet printer, has nozzle unit installed on the spray path of the fluid, which sprays the fluid to the substrate, KR1552433-B1, 11 September 2015 (to Enjet and Univ Singkyunkwan)
14. Hideyuki F et al. (2015) Electrostatic screen printing apparatus for coating powder on target object, e.g., foodstuff, has autorotation mechanism that is arranged to rotate rubbing sponge, and revolution mechanism that is provided to revolve rubbing sponge, WO2016002642-A1, 26 June 2015 (to Hitachi)
15. Comiskey B, Albert JD et al (1998) An electrophoretic ink for all-printed reflective electronic displays. Nature 394:253–255
16. Blankenbach K, Schmoll A et al (2008) Novel highly reflective and bistable electrowetting displays. J Soc Inf Disp 16(2):237–244
17. Argun AA et al (2004) Multicolored electrochromism polymers: structures and devices. Chem Mater 16:4401–4412
18. Shui L, Hayes RA et al (2014) Microfluidics for electronic paper-like displays. Lab Chip 14:2374–2384
19. Heikenfeld J, Drzaic P et al (2011) A critical review of the present and future prospects for electronic paper. J Soc Inf Disp 19:129–156
20. Kao WC, Liu CH et al (2016) Towards video display on electronic papers. J Disp Technol 12:129–135
21. Strubbe F, Vanbrabant PJM et al (2011) In-plane electrophoresis in nonpolar liquids: Measurements and simulations. Colloids Surf A Physicochem Eng Asp 376:89–96
22. Electrocoating (2002) A guide book for finishes. Electrocoat Association, Hanser Gardner, Cincinnati
23. Goud RF (1973) Electrodeposition of coatings, Advances in chemistry series, vol 119. American Chemical Society, Washington, DC, pp i–vi
24. Lambourne R (1999) In: Strivens TA (ed) Paint and surface coatings. Woodhead, Cambridge
25. Yeates RL (1996) Electropainting: a survey of principles and practice covering paint formulation and application. Robert Draper, Sevenoaks
26. Liberto N (2003) User's guide to powder coating. Society of Manufacturing Engineers, Dearborn, MI
27. Barringer SA, Sumonsiri N (2015) Electrostatic coating technology for food processing. Annu Rev Food Sci Technol 6:157–169
28. Luo Y, Zhu J, Ma Y, Zhang H (2008) Dry coating, a novel coating technology for solid pharmaceutical dosage forms. Int J Pharm 358:16–22
29. Halim F, Barringer SA (2007) Electrostatic adhesion in food. J Electrost 65:168–173
30. Gurrappa I, Binder L (2008) Electrodeposition of nanostructured coatings and their characterization: a review. Sci Technol Adv Mater 9:043001
31. Steckl AJ, Heikenfeld JC (2011) Display capable electrowetting light valve, US Patent 7872790-B2, 18 January 2011 (to the Univ Cincinnatti)
32. Heikenfeld J, Steckl AJ (2005) Hight-transmission electrowetting light valves. Appl Phys Lett 86:151121
33. Shendero A (2004) Electrostatic actuators for microfluidics and methods of using the same, US Patent 6,773,566 B2, 10 August 2004 (to Nanolytics)

34. Kuiper S et al. (2004) Zoom lens, WO 2004/038480 A1, 6 May 2004 (to Philips Electronics)
35. Parker KR (2007) Electrical operation of electrostatic precipitators. The Institution of Engineering and Technology, London
36. Parker KR (ed) (1977) Applied electrostatic precipitation. Blackie Academic, London UK
37. Mainelis G, Grinshpun SA et al (1999) Collection of airborne microorganisms by electrostatic precipitation. Aerosol Sci Technol 30:127–144
38. Mizuno A (2000) Electrostatic precipitation. IEEE Trans Dielect Elect Insulation 7(5):615–624
39. Elahi A, Fowowe T et al (2012) Dynamic electrochemistry in flame plasma electrolyte. Angew Chem Int Ed 51:6350–6355
40. Chan TL, Lee PS et al (1981) Diesel-particulate collection for biological testing. Comparison of electrostatic precipitation and filtration. Environ Sci Technol 15(1):89–93
41. Roux JM, Sarda-Estève R et al (2016) Development of a new portable air sampler based on electrostatic precipitation. Environ Sci Pollut R 23(9):8175–8183
42. Foat TG et al (2016) A prototype personal aerosol sampler based on electrostatic precipitation and electrowetting-on-dielectric actuation of droplets. J Aerosol Sci 95:43–53
43. Ma Z et al (2016) Development of an integrated microfluidic electrostatic sampler for bioaerosol. J Aerosol Sci 95:84–94
44. Sueto T, Ota Y et al (2013) Suppression of dust adhesion on a concentrator photovoltaic module using an anti-soiling photocatalytic coating. Sol Energy 97:414–417
45. Xue M, Yan G et al (2012) Electrostatic separation for recycling conductors, semiconductors, and nonconductors from electronic waste. Environ Sci Technol 46:10556–10563
46. Chuetor S, Luque R et al (2015) Innovative combined dry fractionation technologies for rice straw valorization to biofuels. Green Chem 17:926–936
47. Shuman TR, Mason G et al (2016) Low-energy input continuous flow rapid pre-concentration of microalgae through electro-coagulation-flocculation. Chem Eng J 297:97–105
48. Oren Y (2008) Capacitive deionization (CDI) for desalinization and water treatment—past present and future (a review). Desalinization 22:10–29
49. Li J, Xu Z et al (2007) Application of corona discharge and electrostatic force to separate metals and nonmetals from crushed particles of waste printed circuit boards. J Electrost 65:233–238
50. Bryan T, Macleod T et al. (2015) Innovative electrostatic adhesion technologies, NASA Marshall space flight center. http://ntrs.nasa.gov/archive/nasa/casi.ntrs.nasa.gov/20150021399.pdf. Accessed 22 August 2016
51. Shintake J, Rosset S et al (2016) Versatile soft grippers with intrinsic electroadhesion based on multifunctional polymer actuators. Adv Mater 28:231–238

Chapter 14
Instrumentation

Contents

14.1 Measuring Charge, Potential, and Field

The change underwent by electrostatics in the last few decades was made possible by recent experimental developments, both radical and incremental. This is an area where the impact of new experimental tools was paramount, while theory did not contribute significantly.

The two most important quantities in electrostatics are electric charge and electric potential. The product of charge and potential difference is the energy required to transfer charge to a higher potential. Thus, charge and potential are all the electrical data needed for a thermodynamic analysis of electrostatic phenomena: the first is an extensive property while the other is intensive. Charge and potential can now be measured with great accuracy and precision as shown in the forthcoming sections.

On the other hand, mechanical analysis of electrostatic phenomena is rendered easier by using the electric field that is the gradient of potential, especially when one desires to find the electric forces on a charged body and the resultant of all acting

F. Galembeck, T.A.L. Burgo, *Chemical Electrostatics*,
DOI 10.1007/978-3-319-52374-3_14

forces. This is essential in the analysis of their trajectories, as in mass spectrometry. DC field measurement is not as easily done as DC potential measurements but its calculation may be straightforward, depending on the overall geometry of the system of interest.

Measurements of surface potential and charge or electric field [1, 2] are performed by using induction probes, field mills, and electrostatic probes. All of them are based on the capacitive coupling of a conductive surface bearing the surface charge density to the inducing electrostatic electric field.

An example of an induction probe [3] is shown in Fig. 14.1, where a small metal sphere is connected to one end of a segment of coaxial cable. The external conductor of the cable is grounded, and it is encased in a glass tubing that acts as a support and a mounting site. The other end of the coaxial cable is connected to the input of an operational amplifier with high input impedance and suitable signal filtering. Induction probes measure the induced charge by integration of the current, and they require previous cancellation of the charge on the sensor, by exposing it to a zero voltage reference. Only short-time measurements are reliable, due to adverse effects of leakage currents and changing electric patterns in the environment. This problem is solved in field mills by using a mechanical chopper that provides a zero reference for the sensing surface, with a chosen frequency. The AC electric field seen internally by the sensor is easier to amplify with low noise than a DC field. Direct read-

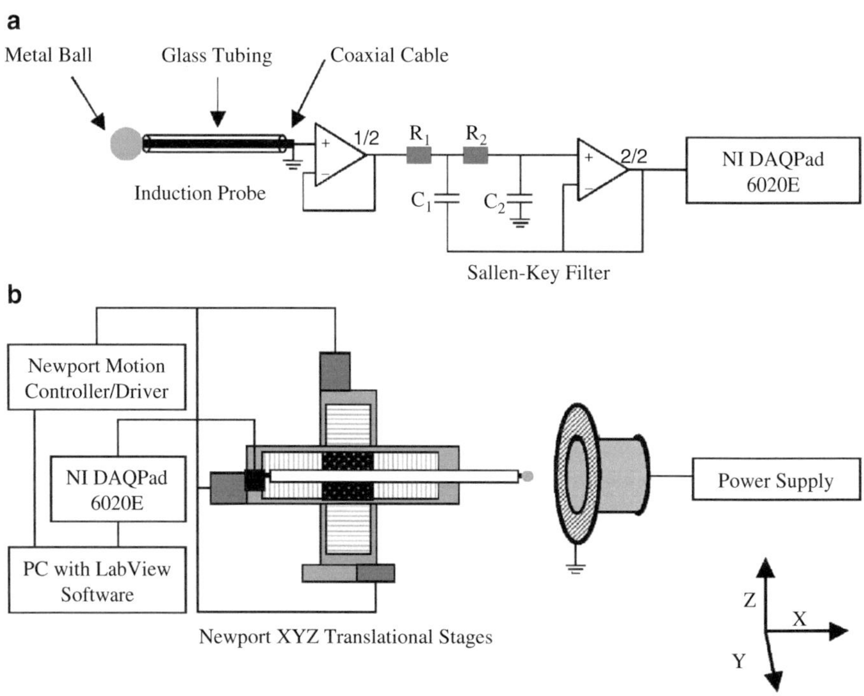

Fig. 14.1 (**a**) A simple induction probe and (**b**) schematic representation the experimental setup. Reprinted with permission from Ref. [3]

ing in volts is possible but it depends on the probe to surface distance and it is impossible, for instance, when the field produced by a cloud is measured in instruments placed on the ground.

A *charge amplifier* or *charge-to-voltage* converter is described by Blackburn [4]. Its output is a voltage that increases linearly with time. Commercial piezoelectric force and pressure sensors are available with integrated charge amplifiers that provide a low-impedance voltage output.

Electrostatic probes like the Kelvin probe (Fig. 14.2) are fed by an integrated high voltage source driving the probe potential in order to cancel the electric field between the surface and the probe. They are well suited for non-contact measurement of potential in adjacent solid or liquid surfaces but precise measurements can only be done under short distances between probe and sample, typically 2 mm. Beyond, the probe-to-ground distance has also to be taken into account to prevent large systematic error [5].

There are many charge detection methods, and a report on various macroscopic methods suitable for large quantity and volume investigations, in research and industrial applications was organized by Noras [6]. They are used in different groups of instruments: those built to satisfy some specific need of the interested researchers, general-purpose instruments used in research and those providing information relevant to safety standards. This chapter is not a comprehensive review but it concentrates in some relevant cases, based on their impact in current research on electrostatic charging mechanisms, patterns, and consequences.

14.1.1 Comparative Advantages of Charge Detection

Detection of charged species by electrical measurements is extremely sensitive, compared to any analytical of physical methods based on weight or on other physical properties, including the detection of electromagnetic radiation. This is demonstrated by mass spectrometry, electron spectroscopy, and electroanalytical techniques that are ultimately based on the detection of charge. A spectacular example of the superiority of charge detection techniques is the current ascent of cryo-electron microscopy as a tool for structural determination of proteins and other complex biological structures [7]. This technique now competes advantageously with the hitherto standard electron diffraction but with a major difference in the cost of the required equipment: cryomicroscopy is done using state-of-the-art electron microscopes that nevertheless cost approximately 1/100th of the cost of a synchrotron, and it requires much smaller amounts of sample than the competing X-ray diffraction dispensing with the need for crystallization.

The advantages of charge detection derive from two factors. One is the accuracy and precision achieved in the measurement of electrical magnitudes, especially the electric potential, current, frequency, and capacitance. The other is the elementary charge (1.6×10^{-19} C) that is also expressed in the magnitude of the Faraday constant, 96,485 C/mole [8] of individual positive or negative ions, or electrons.

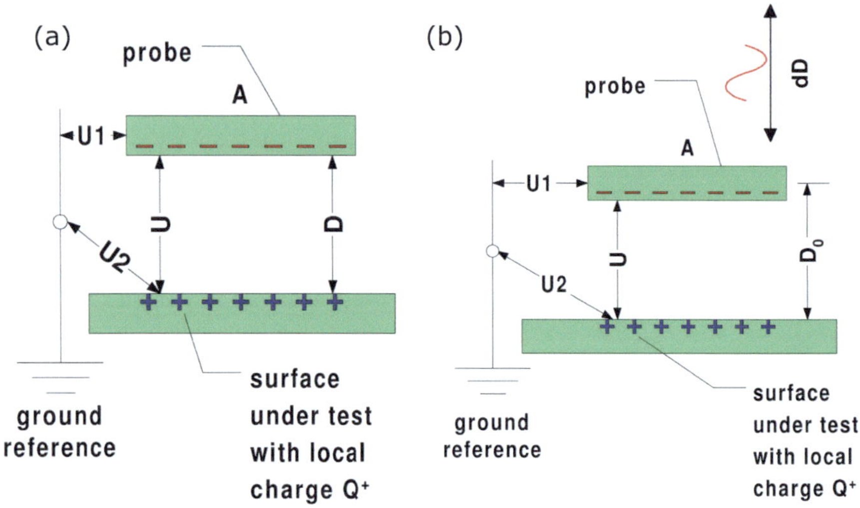

Fig. 14.2 (**a**) A parallel-plate capacitor and (**b**) the schematic description of a Kelvin electrode. Reprinted with permission from Ref. [5]

Simple lab-bench charge meters detect 10^{-9} C, or 10^{-14} moles of ions. This amount of monovalent sodium ions weights 2.3×10^{-13} g that is way below the sensitivity of microbalances. Moreover, a 10^{-9} C point charge produces 2.25×10^6 V at 2-mm distance, which is a very high voltage. Accordingly, 10^{-15} C that is the charge of 6000 sodium ions only creates 2.25 V at 2 mm, a voltage well within the capabilities of simple meters, was not for the need to use a meter with infinite input resistance and low capacitance. These limitations are overcome by using Kelvin electrodes or probes with the associated circuitry.

14.2 Charge Measurement and the Faraday Cup

The Faraday cup is the most often used instrument for charge measurement, in recent literature. It has been used in aerosol analyzers [9], radiation and particle dosimetry [10], mass spectrometry, high-energy physics [11], high intensity proton injection equipment [12], development of high-intensity high-charge state laser ion sources [13], tribocharging of powders and pharmaceutical granules [14], and a host of other instruments and systems. These devices can be built forming sets or arrays: two Faraday cup sensors were used as part of the integrated set of sensors used in the WIND spacecraft solar wind experiment [15], and Faraday cup arrays are built for measurements with spatial resolution [16].

The respective IUPAC definition in the Gold Book [1] is: “A hollow collector, open at one end and closed at the other, used to collect beams of ions.” This definition reflects the main current use of Faraday cups that is measuring the current in a beam of charged particles. In this case, it consists of a conducting metallic chamber or cup, which intercepts a particle beam. An electrical lead is attached to a terminal of an ammeter while the other is connected to ground. Alternatively, a voltmeter or oscilloscope reads the voltage across a resistor inserted in the conducting lead connecting the cup to ground. The importance of Faraday cups to particle physics led to the establishment of the Faraday Cup Award [17] that is given every 2 years, "for an outstanding contribution to the development of an innovative beam diagnostic instrument of proven workability."

A simple version for measuring charge in solids and liquids is built by mounting two concentric pieces of metal tubing, isolated from each other as in Fig. 14.3. The inner cylinder of the cup may have its lower end closed to hold solids or liquids, or open for flow-through measurements on charged material. The outer cylinder is grounded and the inner cylinder is connected to an electrometer for charge measurement but also for potential, current, and other desired measurements, depending on the electrometer used. Time-resolved measurements are acquired using the memory in the electrometer or by connecting the electrometer output to a data acquisition board. Standard equipment allows measurements with resolution in the millisecond range.

More complex design has been used in especial cases, as in the measurement of excess charge in water samples [18].

Handling of Faraday cups requires utmost care because many surrounding disturbances may add or subtract from the stored charge. The apparatus has to be assembled carefully, in a clean environment and it should be grounded prior to use. The experimenter should always recall that grounding capacitive systems is never instantaneous and the required time for reaching the ground voltage depends on the system time constant. Moreover, the experimenter needs to be fully aware about how easily charge is produced when any object or liquid contacts any other object. Just touching the Faraday cup with any surface may well impart or withdraw charge from it. The same happens with dust and aerosol particles.

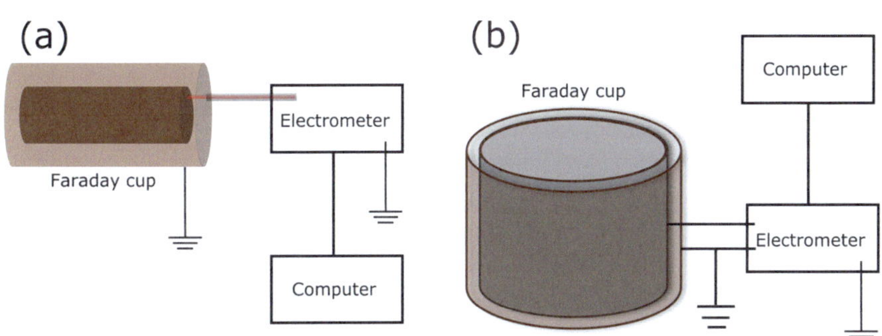

Fig. 14.3 Different Faraday cup designs. (**a**) Faraday cup used for aerosol measurements and (**b**) for the measurement of charge on condensed phases, liquid or solid

14.2.1 *Faraday Cups and Hygroelectricity*

In principle, any metals can be used to make Faraday cups, but in practice the measurements may be affected by specific adsorption or chemical reaction of some contacting substance on the metal surface. Indeed, one of the first evidences of water vapor ability to impart charge to a surface was obtained by using Faraday cups made of chrome-plated brass, as described in Chap. 6, on hygroelectricity. For this reason, metals with low ability for selective adsorption of H^+ or OH^- ions are preferred, like copper that does not show significant charge under variable humidity. Faraday cups used in highly sensitive instruments are often made out of gold [19], or gold plated.

The charge measured in a Faraday cup is actually the algebraic sum of all excess charges within the system used. In some cases, a small charge is measured in a bipolar system that carries separated particles with positive or negative charges, due to mutual cancellation. For instance, when an aerosol flows through a Faraday cup and time-resolved charge measurement is made, large positive and negative peaks are observed, as shown in Fig. 6.20. However, the charge of the accumulated aerosol is the result of cancellation of the separated opposite charges in the aerosol.

14.3 Electric Potential: The Kelvin Probe

Most important recent findings on electrostatics were made possible by the availability of practical, robust, and precise Kelvin probes and the associated measuring circuitry. This technique is named after its inventor [20], and it was perfected [5, 21] along the years. Currently, the probe is a vibrating electrode that forms a parallel-plate capacitor with the surface that is under examination, following Zisman [21].

The Kelvin Probe is a noninvasive analytical tool extremely sensitive to changes in the upper atomic layers of a sample, such as those caused by adsorption, absorption, deposition, wear, corrosion, and atomic displacement. It can detect less than one-thousandth of an adsorbed layer, according to results presented in various chapters in this book.

Even so, surface potential measurements have been often absent from books and reviews on surface analysis, with the exception of the colloid and interface scientists studying monolayers on liquids [22, 23]. However, they are undergoing an impressive renaissance in this century that was largely stimulated by the Kelvin probe force microscopes [24]. Recent progress is due to advances in hardware design and signal processing technology that improved the resolution of the instruments. It can now be used in many different types of environment, including vacuum. Most relevant for the objectives of this book, the Kelvin probe can be used in almost any laboratory as well built and open air environments, under changing pressure and humidity. Improvements have been made to the spatial resolution of this technique that is now applied to macroscopic samples with a few-millimeters spatial resolution as well as using various modes of the scanning probe microscopes, with a resolution in the 10–30 nm range.

Bibliometric data show the impressive and solid growth of the importance of the Kelvin electrode. In a search done on Web of Science in September 2016, using the search terms "Kelvin electrode" OR "Kelvin probe" NOT "microscopy" 1120 publications were reported for all years (1900–2016), growing from 26 in 1997 to 82 in 2013, the peak year. The number of citations in the same period grew from little more than 200 to >2000 in 2016 and the number of average citations per item is 18, giving an h-index equal to 64. Main users are in the following areas: materials science, applied physics, condensed matter physics, physical chemistry, electrochemistry, and related areas.

As for the nanometric scale, the corresponding numbers for the search terms "Kelvin electrode" OR "Kelvin probe" AND "microscopy" are: 2030 publications, rising from less than 10 in 1997 to 216 in 2016. In the same period, the number of citations rose by two orders of magnitude, reaching nearly 5000 in 2015, with 17 average citations per item, h-index 80. Main users are in the same areas that are important for the macroscopic probes but with the addition of "nanoscience and nanotechnology."

Kelvin probes responds to the work function of metals and semiconductors including organic conducting polymers. They have provided relevant information in sensor research [25] and allowed to understand how a metal oxide gas sensor works [26]. Other applications are in the study of organic electroluminescent devices, e.g., to understand enhanced carrier injection [27].

Kelvin probe measurements on insulators allow the identification of domains with fixed charge, and they are an essential tool for measuring potential and therefrom calculating charge on insulating solids and liquids. These probes held a major role in showing the coexistence of domains with opposite charge adjacent to each other, in common thermoplastics and rubbers.

Most important, charge measurements made using Kelvin probes show reasonable agreement with results from Faraday cup measurements. This was experimentally verified by doing the two types of measurements on tribocharged samples of polytetrafluoroethylene (PTFE), poli(methyl methacrylate) (PMMA), and polyethylene, shown in Table 14.1. The two first were charged by shearing with a disc of polyethylene foam (PE) while the latter was charged by rolling glass spheres. Electrostatic potential maps on each tribocharged sample were determined using a macroscopic scanning Kelvin probe, followed by measuring its charge in a Faraday cup. The results show that charge obtained by both methods agrees within 10% or

Table 14.1 Charge measured with a Faraday cup or counting the contribution of all pixels on electrostatic maps recorded with a Kelvin probe [28]

Material	Faraday cup/Coulomb	Superposition principle/Coulomb
PTFE × PE (foam)	5.51×10^{-9}	6.01×10^{-9}
PMMA × PE (foam)	6.78×10^{-9}	6.61×10^{-9}
PE (film) × glass beads	-3.25×10^{-8}	-3.07×10^{-8}

better. This is remarkable, considering that the two measurements are completely independent (except for the sample used) and also that the potential measured on each pixel in the potential map is an average value.

Usage of the Kelvin electrode is not yet widespread in biomedical sciences, even though specialized equipment has been built with this purpose [29]. This is a multi-tip scanning Kelvin probe, which can measure changes in biological surface potential ΔV to within 2 mV while monitoring displacement shorter than 1 μm. The system permits long-term (>48 h) "active" tracking of the displacement and biopotentials developed along and around a plant shoot in response to an environmental stimulus, like differential illumination (phototropism) or changes in orientation (gravitropism).

14.3.1 New Developments

New instrumental developments are appearing continuously, and we may expect that these will expedite discovery relevant to the whole area of electrostatics. For instance, a recent paper [30] reports the implementation of a three-dimensional mapping routine for probing solid–liquid interfaces using frequency modulation atomic force microscopy. This enables fast and flexible data acquisition of up to 20 channels simultaneously, and it is extendable to Kelvin probe force microscopy.

Another development that further expands the scope of the Kelvin probe is a setup for the measurement of local surface photovoltage spectra within a Kelvin microscope operated under ultrahigh vacuum conditions. It allows spectra to be recorded as a function of the wavelength of the illuminating light [31].

14.4 Commercially Available Equipment

There are many types of meters suitable for electrostatic measurements and a bigger diversity of the names used by different makers. Most suppliers of these instruments are specialized companies that are not familiar even to persons involved with instrumentation for electrodynamic, electromagnetic, and electrochemical measurements. In the following short list, the different instrument designations used by the suppliers are represented, even if they refer to rather similar instruments:

Kelvin probes, vibrating capacitor meters: KP Technology Ltd. (UK); Monroe Electronics Inc. (Lyndonville, NY); Trek Inc., (Lockport, NY).

Kelvin probe microscopes: these are supplied as attachments for atomic force microscopes, by most companies selling these instruments. In the earlier years of this technique, different names where used for essentially the same thing: for instance, the Topometrix company called it "scanning electron potential microscopy" (SEPM) while Digital used "electric force microscopy" (EFM). For Topometrix, EFM was another procedure for measuring charge, not potential.

Static meters, surface voltmeters, air ion counters: AlphaLab Inc. (Salt Lake City, UT); Electro-Tech Systems, Inc. (Glenside, PA); Kasuga Denki Inc. (Kanagawa, Japan); Keyence (New Tech Park, Singapore); PCE Instruments UK Ltd. (Southampton, UK); Wolfgang Warmbier GmbH & Co. KG (Hilzingen, Germany)

Electrostatic locators, detectors, or monitors: LessEMF.com (Latham, NY), Motion Industries (Birmingham, AL); Static Clean International (North Billerica MA);

Faraday cups and nanocoulombmeters: Electro-Tech Systems, Inc. (Glenside, PA); Monroe Electronics Inc. (Lyndonville, NY); Static Clean International (North Billerica, MA);

DC Field meters: (including AC field meters that may operate at low frequencies, down to 5 Hz): LessEMF.com (Latham, NY), Monroe Electronics Inc. (Lyndonville, NY); PCE Instruments UK Ltd. (Southampton, UK)

Electrostatic safety and control equipment, devices for elimination of static charging: Botron Co. (Phoenix, AZ); Exair Co, (Cincinatti, OH), MSC Industrial Supply (Melville, NY); Motion Industries (Birmingham, AL); Static Clean International (North Billerica, MA); ElectroStatics Inc. (Hatfield, PA).

This list is not intended to be comprehensive and the authors apologize for any omissions. This is included in this book with the sole intention to help beginners who are not familiar with the suppliers. In the authors' experience, most of the mentioned suppliers are not known to persons who have not done experimental research in this area or who are not involved with safety problems, in industry. Moreover, these products do not appear in the catalogs of broad scope laboratory equipment companies, with very few exceptions.

14.5 Apparatus for Specific Measurements

Many specialized instruments have been developed to answer specific needs of experimenters under diverse conditions. This section provides a glimpse on some instruments that are especially interesting, for their ingenuity or for the importance of results provided by them.

Scientists from the *Kennedy Space Center's Electrostatics and Surface Physics Lab* at NASA led by Dr. Carlos Calle are developing an electrostatic analyzer called Wheel Electrostatic Spectrometer (WES) as a surveying tool to be incorporated into a planetary rover wheel [32]. Electrostatic sensors with various cover insulators are embedded into a prototype wheel to analyze how these insulators charge against other materials. The sensor cover insulators made of polytetrafluorethylene (PTFE – Teflon), poly(methyl methacrylate) (PMMA), glass epoxy laminate (G10), and polycarbonate were strategically chosen based on their respective locations in the triboelectric series. Since these materials cover the triboelectric series from top to bottom, they will charge differently when rolling over the Martian regolith, creating a *charge spectrum* [32]. The signal from each sensor head is sent to a one nF capacitor and the voltage generated is amplified and sent for analysis [32].

Another topic that requires special apparatus is the study of the interplay between friction and tribocharge. Budakian and Putterman [33] developed their own electrostatic apparatus that basically consists of a metal tip mounted on the end of a cantilever, as shown in Fig. 14.4. While the vertical displacement of the cantilever is used to apply a normal force, the lateral deflection measures the frictional force. Surface charge is evaluated by measuring the capacitance between a metal strip that is attached to each cantilever and a grounded metal chopper placed between the back of the dielectric sample and a metal plate [33]. These authors measured a charge density of 10^8 unit charges/mm^2 on nylon surface, after being scanned by a graphite tip, showing a strong correlation between macroscopic friction and triboelectrification of materials.

An important feature of tribological experiments, which is seldom used to investigate the triboelectrification of materials, is the flow of charged species at the interface, called *tribocurrent* [34, 35]. This small electric current between two surfaces under relative motion is the result of the different electron work functions for metal–metal interfaces and of complex mechanisms for metal–insulator and insulator–insulator interfaces.

Burgo and Erdemir used a ball-on-disk friction test geometry to record the macroscopic friction force and simultaneously measure the current generated at the metal–insulator interface [36]. As schematically shown in Fig. 14.5a, friction and tribocurrent experiments were conducted on a CSM high vacuum tribometer with a ball-on-disk contact geometry with a Keithley (6514) electrometer attached. The CSM tribometer metal chamber acts as a Faraday cage, preventing disturbances due

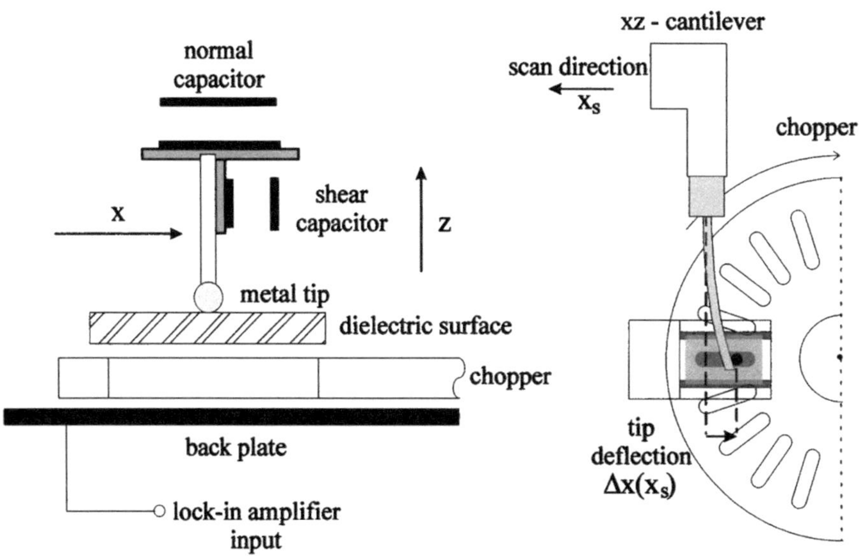

Fig. 14.4 Cantilever and integrated charge measurement apparatus. Reprinted with permission from Ref. [33]

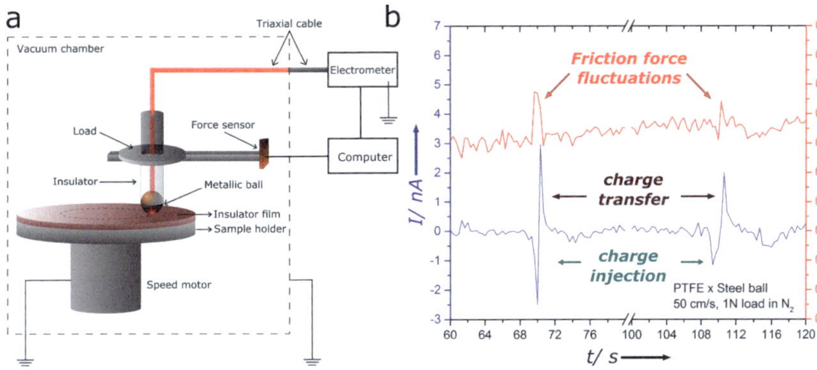

Fig. 14.5 (**a**) Simultaneous measurement of the tribocurrent and macroscopic friction force using the ball-on-disc geometry. (**b**) Friction force fluctuations and tribocurrent generation at the metal–PTFE interface. Reprinted with permission from Ref. [36]

to external electrical fields. The apparatus records both friction force and tribocurrent, under different conditions of load and speed [36] and the results show that friction force fluctuations (stick-slip) and bipolar charging events at metal–insulator interfaces are simultaneous (Fig. 14.5b). During random events of force maxima, charges are exchanged in both directions, from the metal to the insulator and in the opposite direction. The magnitude of charged species exchanged across the interface is highly dependent on the surrounding atmosphere. Besides, mechanical contact increases the pull-off force fifteen-fold, producing a resilient electrostatic adhesion [36].

Whitesides' group at Harvard developed many special tools for studying contact electrification at metal–insulator interfaces. In 2003, they described an analytical system for in situ measurement of the charge developed by contact electrification when a ferromagnetic sphere rolls on the surface of a polymer [37]. This apparatus yields data on polymer surface charging by contact electrification and its kinetics, but without physical contact between the charged sphere and the measuring electrode [37].

Atmospheric Pressure Interface Time-of-Flight Mass Spectrometer (APi-TOF, Tofwerk AG) is a representative of instruments designed to measure atmospheric ion composition. The APi-TOF consists of a time-of-flight mass spectrometer (TOF) coupled to an atmospheric pressure interface (APi) that drives ions from atmospheric pressure to the TOF while pumping away the gas. The APi should not be confused with atmospheric pressure ionization, as the APi-TOF does not by default contain any ionization method. The apparatus measures atmospherically relevant compounds, such as charged $(H_2SO_4)_m(NH_3)_nHSO_4^-$ clusters, in the laboratory and in the environment [38].

14.6 Conclusion

There is now a large number of instruments for measuring electrostatic charge, potential and associated magnitudes but these are still not familiar to researchers in many areas and even to students. Their presentation to science and engineering students is necessary, contributing to spread up-to-date ideas on electrostatic phenomena. Beyond, new developments are needed and any new procedures that provide faster, more accurate, and precise measurements covering broader ranges of charge and potential in different systems will certainly contribute to new important findings.

As a special note, equipment for electrostatic safety monitoring and risk abatement should become more widespread to avoid the persistent accidents and disasters originated from electrostatic charging.

References

1. Chang JS, Kelly AJ, Crowley JM (1995) Handbook of electrostatic processes. Marcel Dekker, New York
2. Llovera P, Molinié P, Soria A, Quijano A (2009) Measurements of electrostatic potentials and electric fields in some industrial environments. J Electrost 67:457–461
3. Appel MF, McKeachie JR, van der Veer WE, Benter T (2004) Simple induction probe electric field meter for the detection of electrical fields generated by ion-optical electrodes. Rev Sci Instrum 75:2603
4. Blackburn JA (2001) Modern instrumentation for scientists and engineers. Springer, New York, p 204
5. Noras MA (2002) Non-contact surface charge/voltage measurements. Capacitive probe—principle of operation. Trek Application Note 3001, Trek, Lockport.
6. Noras MA (2013) Charge detection methods for dielectrics—overview. Trek Application Note 3005, Trek, Lockport.
7. Sirohi D et al (2016) The 3.8 angstrom resolution cryo-EM structure of Zika virus. Science 352:467–470
8. McNaught AD, Wilkinson A (1997) IUPAC Compendium of chemical terminology (the "Gold Book"), 2nd edn. Blackwell Scientific, Oxford
9. Flagan RC (1998) History of electrical aerosol measurements. Aerosol Sci Technol 28(4):301–380
10. Karger CP, Jaekel O, Palmans H, Kanai T (2010) Dosimetry for ion beam radiotherapy. Phys Med Biol 55:R193–R234
11. Brown KL, Tautfest GW (1956) Faraday-cup monitors for high-energy electron beams. Rev Sci Instrum 27(9):696–702
12. Berezov R et al (2016) High intensity proton injector for facility of antiproton and ion research. Rev Sci Instrum 87:02A705
13. Zhao HY et al (2014) The study towards high intensity high charge state laser ion sources. Rev Sci Instrum 85:02B910
14. Naik S, Sarkar S, Hancock B, Rowland M, Abramov Y, Yu W, Chaudhuri B (2016) An experimental and numerical modeling study of tribocharging in pharmaceutical granular mixtures. Powder Technol 297:211–219

15. Ogilvie KW et al (1995) SWE, a comprehensive plasma instrument for the WIND spacecraft. Space Sci Rev 71:55–77
16. Prokůpek J et al (2014) Development and first experimental tests of Faraday cup array. Rev Sci Instrum 85:013302
17. https://www.faraday-cup.com/index.html, Accessed September 2016.
18. Amin MS, Peterson TF, Zahn M (2006) Advanced Faraday cage measurements of charge and open-circuit voltage using water dielectrics. J Electrostat 64(7–9):424–430
19. Nakayama Y, Sohda Y, Ohta H, Saitou N, Muraki M, Takakuwa M (2004) Electron beam monitoring sensor and electron beam monitoring method. US Patent Appl US20040026627, A1.
20. Kelvin L (1898) Contact electricity of metals. Philos Mag 46:82–120
21. Zisman WA (1932) A new method of measuring contact potential differences in metals. Rev Sci Instrum 3:367–368
22. Oliveira ON, Bonardi C (1997) The surface potential of Langmuir monolayers revisited. Langmuir 13:5920–5924
23. Karakashev SI, Nguyen AV, Miller JD (2008) In: Narayanan R (ed) Interfacial processes and molecular aggregation of surfactants, Advances in polymer science, vol 218, pp 25–55
24. Galembeck F, Costa CAR (2006) Electric scanning probe techniques: Kelvin force microscopy and electric force microscopy. In: Somasundaran P (ed) Encyclopedia of surface and colloid science, vol 3, 2nd edn. CRC Press, Boca Raton, pp 1874–1883
25. Moos R et al (2009) Solid state gas sensor research in Germany—a status report. Sensors 9(6):4323–4365
26. Bârsan N, Weimar U (2003) Understanding the fundamental principles of metal oxide based gas sensors; the example of CO sensing with SnO_2 sensors in the presence of humidity. J Phys Condens Matter 15:R813
27. Appleyard SFJ, Day SR, Pickford RD, Willis MR (2000) Organic electroluminescent devices: enhanced carrier injection using SAM derivatized ITO electrodes. J Mater Chem 10:169–173
28. Burgo TAL (2013) Triboeletrização de Polímeros Dielétricos: Mosaicos Macroscópicos de Carga e seus Efeitos sobre as Forças de Atrito em Interfaces, Ph.D. thesis, Unicamp, Campinas.
29. Baikie ID, Smith PJS, Porterfield DM, Estrup PJ (1999) Multitip scanning bio-Kelvin probe. Rev Sci Instrum 70:1842–1850
30. Söngen H, Nalbach M, Adam H, Kühnle A (2016) Three-dimensional atomic force microscopy mapping at the solid-liquid interface with fast and flexible data acquisition. Rev Sci Instrum 87:063704
31. Streicher F, Sadewasser S, Lux-Steiner MC (2009) Surface photovoltage spectroscopy in a Kelvin probe force microscope under ultrahigh vacuum. Rev Sci Instrum 80:013907
32. Johansen MR, Mackey PJ, Holbert E, Clements JS, Calle CI (2013) Characterizing the performance of the wheel electrostatic spectrometer. Proceedings of the ESA Annual Meeting on Electrostatics, G-Session: Special Session on Electrostatic Developments at NASA, G2, 1–7.
33. Budakian R, Putterman SJ (2000) Correlation between charge transfer and stick-slip friction at a metal-insulator interface. Phys Rev Lett 85:1000–1003
34. Akbulut M, Godfrey Alig AR, Israelachvili J (2006) Triboelectrification between smooth metal surfaces coated with self-assembled monolayers (SAMs). J Phys Chem B 110:22271–22278
35. Escobar JV, Chakravarty A, Putterman SJ (2013) Effect of anodic oxidation of single crystal boron doped diamond on tribocurrent and macroscopic friction force with metals. Diam Relat Mater 36:8–15
36. Burgo TAL, Erdemir A (2014) Bipolar tribocharging signal during friction force fluctuations at metal-insulator interfaces. Angew Chemie Int Ed 53:12101–12105
37. Wiles JA, Grzybowski BA, Winkleman A, Whitesides GM (2003) A tool for studying contact electrification in systems comprising metals and insulating polymers. Anal Chem 75:4859–4867
38. Junninen H, Ehn M, Petäjä T, Luosuvärji L, Kotiaho T, Kostiainen R, Rohner U, Gonin M, Fuhrer K, Kulmala M, Worsnop DR (2010) A high-resolution mass spectrometer to measure atmospheric ion composition. Atmos Meas Tech 3:1039–1053

Chapter 15
Perspectives

Contents

15.1 Current Situation

The latter 20 years witnessed unexpected experimental results acquired using new, powerful instrumentation. Efforts to understand these results gave birth to original views on electrostatic phenomena that have been successfully verified, creating a consistent but complex conceptual framework. Relevant new information has been contributed by many disciplines engaged in the study of matter and life, from atmospheric science to biosciences, passing through chemistry, physics and every branch of engineering.

The distance between the current position of electrostatics and its status less than one century ago may be assessed by checking some key words in well-known reference books. For instance, the index of the classic "Static and Dynamic Electricity" book by W. R. Smythe, whose third edition was published in 1968 [1], does not carry some keywords that now appear frequently in the literature on electrostatic phenomena. A few examples are: ions, atmospheric electricity, and electrostatic discharge. In the preface to the first edition, this author states that all developments

F. Galembeck, T. A. L. Burgo, *Chemical Electrostatics*,
DOI 10.1007/978-3-319-52374-3_15

presented in his book are based on "macroscopic experimental facts rather than on hypothetical microscopic structure of conductors and dielectrics." One of the justifications given by him is: "...the development of the most satisfactory theories, those based on wave mechanics, require mathematical technique with which we have assumed the reader to be unfamiliar when starting this book...". These statements allow the reader to understand why electrostatics did not quite evolve during most of the twentieth century: lack of progress was due to the then prevalent ideas on the power of quantum mechanics to bring theoretical solutions to most problems of physics and all of chemistry. Of course, the achievements of quantum mechanics have been impressive but it did not play a significant role in the recent development of electrostatics. This is not surprising: it is just another demonstration of the limitations of its predictive power [2].

Indeed, recent progress in electrostatics has been strongly dependent on leaving behind widespread but unproven ideas. One is electroneutrality of common matter and pure substances, another is that charging phenomena are always the outcomes of electron transfer or still, as explained in the previous paragraph, the capabilities of quantum mechanics as a source of explanation and eventually of valid predictions for every phenomena involving atoms and molecules.

The current picture may look displeasing to many persons because it does not derive from any broad, general, presumably elegant theory. Quite the opposite, looking at electrostatics is currently more or less like looking at the rain forest: at a distance, it appears as a seemingly uniform mass. As we approach it we see more and more diversified entities that interact intensely. Still closer approach shows entities with complex subdivisions within a myriad of new ones. Moreover, the phenomenological status of electrostatics is also unpleasant to theory-oriented researchers, but the present authors prefer sound and useful phenomenology to beautiful but misleading theory.

Established theory and time-honored mathematical treatments like Coulomb equation, Poisson and Poisson–Boltzmann equation maintain enormous importance and current computational capabilities allow their growing use under many circumstances. However, the Poisson–Boltzmann equation is not always applicable, because real electrified systems are always jumping from one to another non-equilibrium situation. They involve high-energy species that may change fast or may remain trapped indefinitely, depending on the nature, properties and dynamics of the surrounding environments.

Thermodynamic arguments are extremely helpful pointing out trends and revealing driving forces for the transformation of electrified systems but again, these often take place under nonthermalized conditions. Thus, thermodynamic methods may predict final states but omitting important intermediary steps.

Deterministic chaos is obviously frequent in electrostatic phenomena but only two regimes are familiar to most persons. One is the steady-state observed under low values of the electric potential and electric field (that are the driving forces) and the other is violent electrostatic discharge. Intermediate transitions and oscillatory regimes in electrostatic systems have been discovered but only in the limited context of instabilities in microelectromechanical systems, MEMS [3–5].

Electric oscillations are well-known in electrochemical systems [6] and it is likely that spatial and temporal oscillations of the overall electrostatic potential of the system are also present, in phenomena like the formation of Liesegang rings and Belousov–Zhabotinsky reaction. Chaotic electric discharge was observed during the slip-stick peeling of adhesive tape [7] and we may expect to observe oscillations and chaotic behavior in any related system like, from instance, water dripping from a faucet.

Concerning atmospheric electricity, it is now clear that it is largely produced within the atmosphere itself, by complex interfacial phenomena that have been replicated in the laboratory but still in a limited scale. There are now some explanations for many phenomena of atmospheric electricity but we still fall short of reaching two objectives: a viable technology to scavenge energy from the atmosphere and full protection from lightning.

A bright side in the current status of electrostatics is the convergence of conclusions drawn by researchers with highly different background and perspectives. An example is the participation of ionic species in electrostatic charge build-up and dissipation, another is the importance of interfacial phenomena in charge separation.

The role of the atmosphere as a charge reservoir and a medium for charge transfer coupled to mass transfer is not yet widely acknowledged. Another idea that is still restricted to surface scientists and engineers is the potential difference between surfaces of condensed matter and the corresponding bulk phases.

Based on the previous paragraphs, the authors can make suggestions of topics of special interest for those acquiring an interest in this area. Three helpful but not so common topics are: surface science, mechanochemistry/high-energy chemistry and chaos theory or dynamic systems. It seems that knowledge in these areas will continue to play an essential role in the present phase of electrostatics.

Beyond, the ability to create new instruments has been invaluable, not only complex lab machinery but also relatively simple instruments that provide robust results. Some new instruments produce huge amounts of data, but this is not a serious problem in the present age of big-data research. Building instrument networks for monitoring, for instance, atmospheric or ground electric fields for disaster prevention will also produce big-data and perhaps useful conclusions therefrom.

15.2 Fast-Moving Topics

Some research areas related to electrostatics currently show extraordinary activity and they will probably benefit from new basic or fundamental findings. On the other hand, they will probably also reveal intriguing phenomena spurring basic investigation.

15.2.1 Scavenging Energy from the Environment

Energy scavenging is a topic of great current interest, for many reasons. This is partly due to the success of photovoltaics that nowadays materializes the vision of distributed power generation done on the roofs and windows of houses.

An important driving force for energy scavenging is the need to power autonomous sensors that currently play an ever-increasing role in the environmental, structural, and medical fields. Autonomous energy systems enable measurements in hardly accessible, harsh or hermetic natural or artificial environments including corrosive conditions. Sensing strategies, communication techniques and power management have been discussed in the literature, as well as the design guidelines for practical systems and proposals for applications [8, 9].

Water droplet ability to acquire and store charge has already been used to build a microfluidic energy scavenging device [10], but collecting energy from cloud and storm electricity does not yet appear in the literature.

Piezo and triboelectric transducers are the most frequently used for energy scavenging and they have now reached the nanoscale [11]. Other different types have been described, as for instance: a dielectric elastomeric generator (DEG) [12], another based on electrostrictive polymers with electrets [13] and electrostatic converters of mechanical energy derived from radioisotope decay [14].

15.2.2 Energy Scavenging and Human Health Care

Power supplies based on energy scavenging are attractive alternatives to batteries used in implantable devices inside the human body, whenever it is impossible to use wires. They are also used to power body sensor networks [15] and wearable devices [16–18] whose dissemination will probably grow considerably.

15.3 New Perspectives

15.3.1 Electrostatics, Chaos and Disaster Prevention

A complicating factor in ESD prevention is the possibility to have chaotic-deterministic phenomena taking place under different situations. Reports on oscillation and chaotic behavior in systems subjected to electrostatic forces where previously mentioned in this chapter.

The early Kelvin measurements on atmospheric electricity already showed that the atmospheric electric field is a highly variable quantity on different timescales [19] that is suggestive of a fractal behavior. Field mill measurements are currently used to produce time series that contributed to the identification of maximum electric field mill values as a lightning predictor that contributes skill in forecasting the occurrence of lightning [20]. Interestingly, three of four predictors identified are sensitive to moisture, whose role in charge build-up in various systems was presented in Chap. 6 and other parts of this book.

15.3.2 Electrophysiology and Electrotherapy

This book is certainly not the place to review this topic but it is not possible to evade the following questions: why aren't procedures based on supplying electricity to living bodies commonplace? Why do they appear to be so irrelevant, compared to procedures based on drugs? This question arises because electricity plays crucial roles in living bodies and thus we could expect it to be the basis for some therapies.

Judging from the scientific literature databases, electrophysiology is a mature scientific discipline. However, electrotherapy is often quoted as ineffective and its use is not recommended in medical papers that are highly cited in the literature [21–23]. On the other hand, meta-analysis of research on its effectiveness in some conditions shows positive results [24], but cautiously.

The greatest success in the medical use of electricity seems to be the pacemakers [25] whose scope is being enlarged, following the authors of Reference [18]: "In the 21st century pacemaker developments are no longer solely about reducing mortality but improving morbidity." Perhaps, some analogues to the pacemaker will be created, thanks to the recent learnings on electrical patterns and on the electrical properties of water as an active agent is the production of electricity and its storage.

15.3.3 X-Ray Sources

Triboelectric production of X-Rays progressed quickly from the basic experiments [7] to the first portable XRFluorescence analyzer, named Watson®. According to information disclosed by the company that has been developing this technology, this is the lowest cost XRF instrument on the market [26] and it has performed well in application tests. Success along this line will probably stimulate research on other practical applications, in triboluminescence and related areas.

15.4 Dissemination

New science becomes important when it is disseminated and is part of the education of engineers, researchers and the lay men.

Throughout this book, reference has been made to papers dating from many decades with important conclusions opposite to some prevailing paradigms. Unfortunately, much of this information was largely ignored.

For this reason, a great effort should be made to examine critically the new information and the new electrostatic paradigms, followed by the dissemination of those that are validated. Doing this will spread a richer view on the material world followed by new tools, techniques and products that will make human life safer and will probably contribute to sustainability.

References

1. Smythe WR (1968) Static and dynamic electricity. McGraw-Hill, New York
2. Chibbaro S et al (2014) Chapter 6 Quantum mechanics, its classical limit and its relation to Chemistry. In: Reductionism, emergence and levels of reality, Springer International Publishing
3. Peano F, Coppa G et al (2006) Nonlinear oscillations in a MEMS energy scavenger. Math Comput Model 43:1412–1423
4. De SK, Aluru NR (2005) Complex oscillations and chaos in electrostatic microelectromechanical systems under superharmonic excitations. Phys Rev Lett 94:204101
5. Tajaddodianfar F, Yazdi MRH et al (2015) On the chaotic vibrations of electrostatically actuated arch micro/nano resonators: a parametric study. Int J Bifurcation Chaos 25:1550106
6. Sitta E, Nascimento MA et al (2010) Complex kinetics, high frequency oscillations and temperature compensation in the electro-oxidation of ethylene glycol on platinum. Phys Chem Chem Phys 12:15195–15206
7. Camara CG, Escobar JV et al (2008) Correlation between nanosecond X-ray flashes and stick–slip friction in peeling tape. Nature 455:1089–1092
8. Sardini E, Serpelloni M (2009) Passive and self-powered autonomous sensors for remote measurements. Sensors 9:943–960
9. Murillo G, Agust J et al (2015) Self-suspended vibration-driven energy harvesting chip for power density maximization. Smart Mater Struct 24:115027
10. Yildirim E, Kulah H (2012) Electrostatic energy harvesting by droplet-based multi-phase microfluidics. Microfluid Nanofluid 13:107–111
11. Fan FR, Tang W et al (2016) Flexible nanogenerators for energy harvesting and self-powered electronics. Adv Mater 28:4283–4305
12. Mistral J, Beaune M et al (2014) Energy scavenging strain absorber: application to kinetic dielectric elastomer generator. In: Proc SPIE 9056, Electroactive Polymer Actuators and Devices (EAPAD), vol 9056
13. Belhora F, Cottinet PJ et al (2013) Mechano-electrical conversion for harvesting energy with hybridization of electrostrictive polymers and electrets. Sensor Actuat A Phys 201:58–65
14. Xiang W, Qiang Z et al (2014) Radioisotope energy conversion using electrostatic vibrationto-to-electricity converters. ACTA Phys Sin 63(2):28501
15. Yang GZ (ed) (2014) Body sensor networks. Springer, London
16. Thotahewa KMS, Redouté JM et al (2014) A lowpower wearable dual-band wireless body area network system: development and experimental evaluation. IEEE Trans Microw Theory Tech 62(11):2802–2811
17. Perez JA, Leff HMI et al (2015) From wearable sensors to smart implants—toward pervasive and personalised health care. IEEE Trans Biomed Eng 62(12):2750–2762
18. Fowler AG, Moheimani SOR et al (2014) An omnidirectional MEMS ultrasonic energy harvester for implanted devices. J Microelectromech S 23(6):1454–1462
19. Aplin KL, Harrison RG (2013) Lord Kelvin's atmospheric electricity measurements. Hist Geo Space Sci 4(4):83–95
20. Mazany RA, Businger S et al (2002) A lightning prediction index that utilizes GPS integrated precipitable water vapor. Weather Forecasting 17:1034–1047

21. McAlindon TE, Bannuru RR et al (2014) OARSI guidelines for the non-surgical management of knee osteoarthritis. Osteoarthritis Cartilage 22:363–388
22. Gibson JNA, Waddell G (2005) Surgery for degenerative lumbar spondylosis: updated Cochrane review. Spine 30(20):2312–2320
23. van der Heijden GJMG (1999) Shoulder disorders: a state-of-the-art review. Best Pract Res Clin Rheumatol 13(2):287–309
24. Klawansky S, Yeung A et al (1995) Metaanalysis of randomized controlled trials of cranial electrostimulation—efficacy in treating selected psychological and physiological conditions. J Nerv Ment Dis 183(7):478–484
25. Ward C, Henderson S, Metcalfe NH (2013) A short history on pacemakers. Int J Cardiol 169:244–248
26. Tribogenics (2017) Introducing Watson. http://tribogenics.com. Accessed 18 Feb 2017

Index

F. Galembeck, T. A. L. Burgo, *Chemical Electrostatics*,
DOI 10.1007/978-3-319-52374-3

MIX
Papier aus verantwortungsvollen Quellen
Paper from responsible sources
FSC® C105338

If you have any concerns about our products,
you can contact us on
ProductSafety@springernature.com

In case Publisher is established outside the EU,
the EU authorized representative is:
Springer Nature Customer Service Center GmbH
Europaplatz 3, 69115 Heidelberg, Germany

Printed by Libri Plureos GmbH
in Hamburg, Germany